Industrializing the *Corn Belt*

Industrializing

the Corn Belt

Agriculture, Technology, and Environment, 1945–1972

J.L. ANDERSON

Northern
Illinois
University
Press

DeKalb

© 2009 by Northern Illinois University Press

Published by the Northern Illinois University Press,

DeKalb, Illinois   60115

Manufactured in the United States using postconsumer-recycled, acid-free paper.

All Rights Reserved

Design by Shaun Allshouse

Library of Congress Cataloging-in-Publication Data

Anderson, J. L. (Joseph Leslie), 1966–

Industrializing the corn belt : agriculture, technology, and environment, 1945–1972 / J.L. Anderson.

   p.   cm.

Includes bibliographical references and index.

ISBN 978-0-87580-392-0 (clothbound : alk. paper)

1. Agricultural innovations—Iowa—History. 2. Agricultural productivity—Iowa—History. 3. Agriculture—Iowa—History. I. Title. II. Title: Agriculture, technology, and environment, 1945–1972.

S494.5.I5A56 2009

630.9777′09045—dc22

2008030431

For Emma

Contents

Acknowledgments

It is a pleasure to recognize and thank the many people who helped complete this project. Doug Hurt welcomed me to Iowa State University and proved an outstanding teacher and friend who encouraged me to write something worth reading. Pam Riney-Kehrberg is another good friend and an excellent mentor who contributed much to my teaching and scholarship. In the early stages of my research and writing, Hamilton Cravens asked several important questions that improved the quality of my work. Jim Andrews and Jim McCormick read the entire manuscript and provided an informed critique. The anonymous referees for Northern Illinois University Press were exceptionally helpful. I accepted most of their advice and am grateful for their critiques and confidence.

John Staudenmaier, Suzanne Moon, and the anonymous referees for *Technology and Culture* were also very helpful at an earlier stage of the work. Two chapters of this piece contain portions of my earlier articles: chapter 2 is connected with "War on Weeds: Iowa Farmers and Growth Regulater Herbicides," *Technology and Culture* 46(4) (October 2005): 719-44; and chapter 7 is based on "'The Quickest Way Possible': Iowa Farm Families and Tractor-drawn Combines, 1940–1960," *Agricultural History* 76(4) (2002): 669–88. Both are reprinted here with permission.

Iowa State University extension faculty members Bob Hartzler, Ken Holscher, Rich Pope, and Marlin Rice read portions of this book's manuscript and made several valuable suggestions. Fellow students at Iowa State endured and critiqued several seminar papers that were the foundation of this project. Any errors of fact or interpretation are mine.

Skilled archivists and scholars made the work much easier and more pleasant. At the State Historical Society of Iowa, Sharon Avery provided valuable assistance in the Des Moines facility while Mary Bennett and Kevin Knoot helped in Iowa City. The Special Collections staff at Iowa State University's Parks Library, led by Tanya Zanish-Belcher, provided an excellent work environment, advised me on manuscript collections, screened films, and cheerfully re-trieved boxes of manuscripts day after day. Thanks go especially to Michele Christian, Becky Jordan, and Brad Kuennen who were on staff for all or most of the duration of my research. Janet Huggard, Paul Lasly, and Ramona Wierson of the Iowa State University Sociology Department gave me access to the department's files of research reports. Frank Holdmeyer of Farm Progress Company and the staff of *Wallaces Farmer* were kind hosts during several days I spent working at their offices and gave me permis-sion to reproduce photographs from their files. Mike Krzycki at Behlen Manufacturing Company, Neil Dahlstrom of John Deere and Company, and Nicole Poock at Sukup Manufacturing provided valuable photographs and authorized their use. The State Historical Society of Iowa, Inc., in Iowa City provided a generous grant to defray research expenses. Beth Loecke patiently transcribed tapes of numerous interviews.

Ellen Garber at the University of Kansas was a friend from the first day I entered her office. I would not have completed a master's degree or started work toward a doctoral degree without her help. My friends and col-leagues in the museum world were in my mind throughout this project. I wanted to write a text that public historians who interpret agricultural and rural history would find useful.

The professionals at Northern Illinois University Press deserve special thanks. Melody Herr solicited the manuscript and was a consummate pro-fessional. She was an expert guide to the process of transforming a manu-script into a book. Alex Schwartz carried the project forward when Melody accepted a position with another press. Susan Bean, Julia Fauci, and Shaun Allshouse have been kind, patient, and knowledgeable partners. Many people, including Pippa Letsky, worked behind the scenes and deserve thanks for making this book a reality.

I am grateful to the many people who consented to discuss their experi-ences in farming and extension work. Long before this became an aca-demic project, my interest began in conversations with my grandparents. They introduced me to the farm in so many memorable ways. It was a pleasure and a privilege to talk with many people who have lived through a time of tremendous change in production agriculture and to gain a bet-ter understanding of this fascinating period. I hope they will find my ac-count of farming to be an accurate representation of their experiences. Several of these people are no longer living, and I regret that they did not see the finished product.

Over the years my parents have supported me in many different ways. They seemed to know just what to do and what not to do. For all of this and more, thank you.

My wife, Emma, deserves the most thanks. Her encouragement and sup-port mean the world to me. I dedicate this book to her and to our daughters.

Industrializing the Corn Belt

Introduction

In 1970 *National Geographic* informed its readers that they were living through a remarkable "Revolution in American Agriculture." Feature writer Jules Hilliard and photographer James Blair described innovations such as the breeding of hard tomatoes suitable for mechanized harvest, the growing of crops without plowing, and a plan to eradicate the screwworm fly (a livestock parasite) by breeding female flies with sterile males. Modern farmers employed the latest scientific research and industrial techniques —including large-scale production, strict cost accounting, standardization, reliance on expertise, and the substitution of capital for labor to achieve tremendous gains in productivity. The farmer of 1970 fed himself and forty-two other people whereas his counterpart of 1945 managed to feed himself and only eleven others. Urban Americans who read Hilliard's article and examined Blair's photographs no doubt shared the author's amazement as they learned of the revolutionary technology that characterized modern agriculture.[1]

The transformation of both agriculture and the countryside was only partially visible to most Americans of the early 1970s. Many people viewed rural areas merely as scenery to travel through or fly over. These observers saw the continuity of farm production more than the changes. Large tracts of land were still devoted to crops. Farmhouses and barns, icons of American agriculture, were readily visible from the road, though there were fewer than at the end of World War II. Farmers attracted public attention only when politicians were dealing with the cost of farm subsidies and commodity programs, or during election campaigns when politicians ventured into the countryside to pose in front of tractors and hay bales in an attempt to court farm voters. Newspaper headlines and radio and television broadcasts of the 1960s and early 1970s focused on the war in Vietnam, race relations, unrest in cities and on college campuses, trouble in the Middle East, and numerous other issues both foreign and domestic. As long as food prices were low, most people in America's growing cities and suburbs paid little attention to rural America.

But there were signs of change and even trouble in the role of agriculture in public life and consciousness. Several highly publicized incidents involving farm chemicals sparked interest in a business that most people took for granted. The 1958 debate in Congress over the Delaney Clause, an amendment to the 1938 Food, Drug, and Cosmetics Act, focused attention on the nation's food supply. The amendment, named for a New York congressman, prohibited the use of carcinogens as food additives, regardless of the degree of risk to human health posed by their use. Prior to Thanksgiving 1959 a portion of the nation's cranberry crop was sprayed with herbicide before marketing, which led to a "cranberry crisis," a shortage that shocked Americans who assumed their food was both safe and in ample supply. In her 1962 book *Silent Spring*, Rachel Carson raised questions about farm practices and the impact of agricultural chemicals on both ecosystems and human health. These flash points alerted Americans to the changing landscape of agriculture in the postwar period.

In the 1970s there was renewed and focused attention on the ways farming had changed. The *National Geographic* feature article from February 1970 was benign in tone—a mix of marvel over technological wonders, a few suggestions concerning the costs of those wonders, and speculation about the farm of the future. Other commentators during the 1970s were more critical. Social critics and intellectuals such as Wendell Berry, Barry Commoner, and Jim Hightower complained of both practical and moral problems in the new style of agriculture. In 1978, in the heart of the nation's Corn Belt, the *Des Moines Register* published a series of articles titled "Pesticides: A Growing Question" highlighting the negative consequences of this technology. In 1979 James Risser won a Pulitzer Prize for his series of articles in the *Des Moines Register* on the environmental impact of modern agriculture. Even historians such as Sam Bowers Hilliard, Wayne Rasmussen, John Schlebecker, and John Shover paid attention to contemporary agriculture. Shover's book *First Majority, Last Minority: The Transforming of Rural Life in America* (1977) described a "Great Disjuncture" in modern rural life, in which "technology takes over" and capitalist values trump those of traditional rural communities. Shover recognized the benefits of the new technology, but he also called attention to the costs of technological change.[2]

What had happened on American farms that triggered such a sense of ecological and social crisis? And why did these changes in farming occur? The purpose of this book is to explain farmers' roles in changing agricultural production, to describe the technology they adopted, and to show how they transformed the rural landscape. Farmers, simultaneously as producers and as consumers, decided which particular techniques would help them make a living on their farms. An industrial ethos became an article of faith for farmers during the postwar period, as they accepted and demanded new machines and chemicals and used these to meet their needs. While advertisers, agricultural extension experts, scientists at land-grant schools, bankers, and policy makers had a voice in how farmers conducted business, farm families were the ones who allocated resources to invest in new technology and who lived with the anticipated and unanticipated consequences of using it.

By the early 1970s a new suite of technology was in place on American farms. New machines, chemicals, and buildings replaced human hands, older machines, and outmoded buildings. The application of herbicide, insecticide, and fertilizer was standard on almost all Iowa farms, although the type and amounts of chemicals used varied from farm to farm. A growing percentage of farmers used feed additives to cut livestock production costs. Threshing machines, corn pickers, stanchion milking barns, corncribs, and the practices of feeding livestock by hand and storing loose hay were obsolete or on their way to becoming so by the 1970s. Combines, hay balers, forage harvesters, as well as fully mechanized dairy parlors, feedlots, and confinement buildings became the new standards in mechanization. Automation was the rule as farmers realized the full potential of the tractor and electric power.

What environmental and social consequences did farmers confront as they transformed agricultural production? In Iowa, a leading state in feed grain and livestock production, the industrialization of farm production brought accompanying shifts in land use as farmers focused on soybeans and devoted fewer acres to small grains. At the end of World War II there were approximately 205,000 Iowa farms with roughly 21.5 million acres in cultivation. Iowa farmers produced 462 million bushels of corn and approximately 10.9 million hogs. In 1970 there were only 135,000 farms with approximately 20.4 million acres in cultivation, but production was soaring, with Iowa farmers producing 858 million bushels of corn and 17.8 million hogs. Among the fields, pastures, and farmyards, new and altered populations of plants and insects found niches in industrial agriculture. Farmsteads looked different with new types of buildings suited to large-scale production. By 1972 several significant government regulations were in place that restricted how farmers conducted business—concerning, most notably, the application of chemicals to land, crops, and livestock and the management of livestock waste. The revolution in production also forced the issue of who would stay in farming, as farmers faced decisions about how to use new technology to expand or specialize and whether even to quit farming. By 1970 a sizable minority of those who had been farming in 1945 either would not or could not continue to farm. The cost of purchasing or hiring all the new equipment as well as the cost of increasingly expensive and complicated chemicals were too great or too daunting for many farm families.[3]

The focus of this study is Iowa, a state representing one of the most developed agricultural landscapes in the United States, the Corn Belt. Since the late nineteenth century farmers in the region from eastern Ohio to the Great Plains specialized in raising corn and feeding it to livestock. To this day it is a region characterized by family farming, even as many of those families have incorporated in order to protect family assets from the financial disasters that plague farmers during times of low commodity prices, drought, flood, or infestation.[4]

Iowa spans the Missouri and Mississippi watersheds, covers approximately fifty-six thousand square miles, and ranks among the most productive farmland in the world. The state is especially suited for raising corn, with favorable soil types, hot and humid summers, an annual average of around thirty inches of rainfall, and a five-month frost-free growing season. Throughout most of the twentieth century, mixed farms of diverse crops and livestock were predominant in all parts of the state, with corn and livestock as the most common commodities. Local specialties emerged, with northeastern Iowans producing more dairy products and those in the glaciated central and north-central portions raising more grain. Eastern Iowa farmers raised copious grain and fattened livestock, while those in the western and southern sections focused on livestock production. In spite of their differences, before World War II most farms in the state were more similar than different, with production varying by degree more than in kind. In the late 1940s Iowa's percentage of corn acres surpassed that of any other midwestern state, and in the early twenty-first century the state continues to rank among the top corn-producing states. In 1950 Iowa farmers accounted for almost 11 percent of the total value of livestock and livestock products sold in the United States, and in the early 2000s it led the nation in hog production.[5]

The postwar agricultural transformation occurred after a period of relative consistency in agricultural production techniques, which had lasted from around 1900 to the 1940s. By 1900 most Iowa farmers had moved beyond the farm-making phase of agriculture, with most Iowa land cultivated, grazed, or managed in woodlots. Farmers of the mid- to late 1800s had learned that they needed to apply manure to their fields and practice crop rotation, including tame grasses and legumes between crops of corn and small grains, to keep production up. During the nineteenth century, farmers had phased out most of their pioneer buildings, had fenced most of their fields and pastures, and were using implements such as chilled iron plows, grain binders, threshing machines, mowers, and manure spreaders to farm a larger portion of their acres. These practices were common on Iowa farms at the end of the nineteenth century.

There were, of course, many changes in farming and farm life between 1900 and the outbreak of World War II, but farmers' relationship with technology and the ways they grew crops and raised livestock remained much the same as at the turn of the century. Before 1940 the internal combustion engine, used for stationary power in the farmyard and in tractors in the fields, had replaced many but not all horses. Farmers used tractors for heavy work such as plowing and disking or belt power, but they continued to use horses for cultivating, hauling feed, and numerous other jobs. Even though the tractor replaced many animals in the 1920s and 1930s, it did not necessarily alter farm production. Most farmers rarely used implements especially designed for tractors during these decades, with the exception of tractor-drawn plows. To accommodate the change they modified older horse-drawn equipment for use with tractors. More important, because farmers still needed to practice the same type of corn, small-grain, and forage-crop rotation in order to prevent the depletion of soil nutrients and the buildup of weed populations and insect pests, major shifts in farm production were precluded. Even though farmers needed less hay and oats to feed horses after they had adopted tractors, they still needed a multiyear rotation system of different kinds of crops. They simply used the grain and fodder production previously dedicated for draft animals to fatten more hogs and cattle. Farm size, the number of acres cultivated, and the number of farms remained relatively constant from 1900 to World War II, indicating that tractors were not used to enlarge farms during this period. Farmers used tractors to cut operating costs, since that power source only needed to be "fed" when it was being used. Farmers did not utilize the full potential of tractor power until after World War II, when more mechanical and chemical changes became common. The most important change in production during the interwar years was the transition to hybrid seed. Few farmers used hybrid seed when it was introduced commercially in the 1920s, but by 1942 virtually all Iowa farmers were planting all their corn acres with it. Yield gains through the use of hybrid corn partially offset losses through government acreage-reduction programs and helped farmers meet the increased demands of wartime production.[6]

Powerful people and institutions urged farmers to adopt an industrial model in the 1920s and 1930s, but few farm families could afford to make the kinds of changes recommended by outsiders, even if they wanted to do so. They survived by "making do" with what they had and depended on kin and neighbors to meet their needs. Conditions changed as the Great Depression gave way to war. In 1942 Congress promised to support commodity prices and encourage maximum production in order to ensure adequate food and fiber supply for the United States and its allies. Farmers produced at record levels for guaranteed prices and consequently earned enough income that they could look beyond meeting their immediate needs. During the war there was also little opportunity to spend, other than paying down debt. In 1945 farm families who had paid off debts were ready to reinvest in their farm operations. People feared that a new farm depression would come, like the one they had experienced after World War I, but they also prepared for a new world in which technology would help them solve problems.[7]

World War II changed the old systems fundamentally and marked a significant shift in the use of technology to replace human labor and to expand the scale of production. Manpower shortages, high production goals, government-sponsored research, and the social aspirations of a new generation of farmers formed the context in which farmers made significant changes in the methods they used to raise crops and livestock. Farmers in the postwar Midwest decreased production costs by substituting machines for labor, used pesticides to destroy weed and insect pests that were obstacles to high crop yields and livestock gains, fertilized fields with chemicals, installed automated feeding systems, and added feed supplements that accelerated animals' ability to absorb nutrients and calories. Members of farm families could now accomplish within hours or days tasks that used to require days or weeks of work, often with the help of full-time hired men and itinerant laborers. By 1970 farmers had severed many of the ties to the agricultural production techniques of their parents' generation. They created a new physical and social landscape characterized by larger farms, larger herds, altered crop rotation, different plant populations, more specialized production, rural out-migration, and a lifestyle that more closely resembled that of urban and suburban people.[8]

A major development that motivated farmers to substitute machines and chemicals for labor was the increasing cost of labor. The onset of World War II resulted in a massive labor shortage because many men and women served in the military or worked in defense industries. Regular paydays and the prospect of overtime wages were attractive to many people who had previously been content with farm employment. Higher wages in other sectors of the economy put pressure on farmers to pay more for labor. In 1945 a writer for the state report on agriculture noted that, although farm wages that year were the highest ever paid, record labor shortages continued. This problem persisted through the 1960s. Farmers improved wages, but compensation for farmworkers remained far below that for jobs in manufacturing, construction, and truck driving.[9]

The farm labor shortage coincided with changes in educational attainment and career possibilities in the postwar period. High school graduation rates increased during the middle of the twentieth century. Beginning in the 1940s more young people attended and completed high school, which meant they had more options for employment than their peers from a previous generation. In 1940, 72 percent of sixteen- and seventeen-year-old Iowans enrolled in school while 28 percent of eighteen- and nineteen-year-olds enrolled. By 1960, 86 percent of the former group and 48 percent of the latter enrolled in school. Farm boys who expected to farm upon graduation from high school often did so, but a minority of them did not. Thanks to the G.I. Bill, many veterans from farms could afford to attend college, which contributed to the drain of laborers from the countryside. With fewer workers available, farmers who employed hired hands had little choice but to pay more for labor. Although the rate of wage increase in the Corn Belt was not as fast as it was in the South or in cities, the real value of wages for farmworkers increased by 20 percent from 1950 to 1970.[10]

Compounding farmers' problems of rising labor costs was the cost-price squeeze. From the mid-1950s to the 1960s the costs incurred by farmers increased faster than the prices they received for commodities. This situation was especially difficult from 1951 to 1956 when commodity prices dropped almost 23 percent while non-farm prices remained constant. A study done by Iowa State College (ISC—which was renamed Iowa State University [ISU] in 1959) indicated that average farm income in the state declined from $10,247 in 1953 to $7,051 in 1955. Historian Gilbert Fite observed that the cost-price squeeze compelled many families to leave their farms, but it also helped those who were in a solid financial position increase their landholdings or invest in their farms to gain economies of scale and cut labor costs.[11]

Stories of agricultural change have been told in many different ways, but the stories told first and most in depth have been about institutions and people with the most education, since these groups and individuals were the ones who left the most records. This study tells the story from the perspective of the people who raised the crops and livestock, just as Allan Bogue did in his 1963 study of farming in Iowa and Illinois during the nineteenth century. Bogue focused on the farmer "with dirt on his hands and dung on his boots—and the problems and developments that forced him to make decisions about his farm business." But histories like Bogue's that emphasize how people used cows, plows, and sows to make a living have fallen out of favor over the last few decades. A new generation of historians rightly brought innovative and previously neglected perspectives on rural life to the forefront. Most of these new studies do not place production agriculture at the center of the story, however. As historian Robert McMath observed about rural social history, "Crops are grown and they are harvested. But with what, how, and by whom?" The details of agricultural production and what they can tell us about rural people are too often left behind. The chapters that follow provide an in-depth and textured examination of farmers and changes in everyday life on the farm, to inform the broad story of one of the most tremendous productive transformations in human history.[12]

This study is organized in two parts to divide complicated and interrelated technological changes into discrete elements. Part I addresses the rise in the use of chemical insecticide, herbicide, fertilizer, and feed additives such as antibiotics and growth hormones. Iowa farmers were on new ground with chemical farming. Only a small minority of farmers used chemical fertilizers before the war, and the new pesticides, synthetic antibiotics, and growth hormones were products of wartime research and development programs. As a result farmers climbed a steep learning curve with the chemicals, using them and misusing them in combination with traditional cultural practices in order to control weeds and insects, while replacing animal manure with chemical fertilizer. Farmers who used this combination of chemical technology realized tremendous productive gains in both crop and livestock production. They also lived with many unanticipated consequences of their decisions to employ chemicals, including the proliferation of resistant species of plants and insects and the growing concern among the public and government officials about the health and safety implications of farm chemicals.

Part II shows how farmers used developing mechanical technology to assist with the harvest of grain, soybeans, and corn, as well as how they experimented with the architecture of grain storage and automated materials handling. Farmers had a great deal of experience with machines, which allowed them to make a smooth transition to tractor-drawn and self-propelled harvest implements. Cost savings, the benefits of reducing dependence on hired labor, and gaining time to change work cycles or for leisure were critical benefits that came with the use of new machines. While some machines such as tractor-drawn combines and hay balers helped farmers reduce labor costs, other machines such as combines for corn and the buildings and crop dryers needed to store shelled corn rather than ear corn were some of the most expensive items farmers would ever purchase. As the 1970s began, self-propelled combines and fully automated feeding systems were the tools for a minority of farmers, but it was a rapidly growing minority.

The topical organization of this book reflects the fact that farmers made many decisions as they transformed the land and that they made these decisions in response to specific issues or problems related to work, livestock, or crop cycles. In one sense, all these issues are related. For example, farmers who invested in fertilizer and pesticides for the corn crop wanted to obtain maximum returns on their investments and, therefore, decided to invest in corn-harvesting equipment that was more efficient and left less of the crop in the field. But in another sense, each technology was distinct enough that decisions about using it often stood independent of other technology decisions. Fertilizer use did not necessarily influence farmers' decisions about herbicide or haymaking equipment. The organization of this book allows readers to see both the similarities and the differences between farm practices and the trajectories of technological change. Some changes were slow and others were rapid. Some practices raised environmental concerns among farmers and outsiders, and others did not. Each technique helped reshape the landscape in its own way. Taken together we see how each technological innovation complemented the others. In this sense, it is possible to understand the technology that farmers used to adjust to changing times as well as the ways farmers themselves changed the landscape of farming.[13]

This study is not a comprehensive account of all the technological changes on Iowa farms during the postwar period, nor does it include all possible perspectives on the subject. Many innovations such as artificial insemination, changes in genetics and breeding standards, and larger, more powerful, and complicated tractors became common after World War II. These changes and more were important but await future study. The mechanical and chemical changes depicted here tell about how farmers used technology, and why, and provide ample evidence of the role they played in shaping technological and environmental change. Farm policy, the farm credit and lending industry, alternative visions of farming, protest, and gender analysis are also beyond the focus of this book, which tells one of the many possible stories in the history of midwestern agricultural production, technology, and environmental change.

Farm families adopted new technology to meet their needs during a time of tremendous change. Few of them either took up or rejected every technological innovation. Most farmers were selective, including those who finally left agriculture. On occasion agricultural extension professionals, advertisers, manufacturers, and journalists all advised, even pleaded, with farmers to change their behavior. Bankers and other lenders were some of the most powerful players in changing farm practices, particularly during the 1950s and 1960s as memories of the Great Depression faded and federal farm programs became entrenched. Farm debt grew from four billion dollars in the early 1950s to eight billion dollars by the end of the decade and grew by approximately three billion dollars each year of the 1960s. The availability of easy credit facilitated the purchase of new equipment, to be sure, but it was the farmers who came to town to sign the papers and commit to new machines. To say that farmers were leaders in the process of agricultural change in the postwar period is not to deny that other parties were important actors in technological change, but it was the farmers who made the decisions that altered the nature of the midwestern landscape.[14]

The world created by these farmers was far different from the one they had inherited in 1945. The similarities begin and end with the facts that much of the land is still used for agricultural production and much of it is planted to corn. Differences, however, abound. One of the most noteworthy is the lack of people. Except for in the spring and fall it is possible to drive through large sections of rural Iowa without seeing anyone except other travelers on the road. Gone are the days of passing several farmsteads each mile, with people working outdoors in the fields and farmyards. The disappearance of livestock and the diminished diversity are nearly as striking. In 1945 almost every farm family kept a wide variety of livestock, in small lots and pastures, and all visible to passersby. In the twenty-first century, depending on the time of year or the route, it is possible to cross an entire county without seeing any livestock. The animals are still there, but they are concentrated increasingly on fewer farms and are often indoors. Even the fields look different. The old crop rotations of corn, small grain, and hay are gone, as well as other crops such as flax. Today corn and soybeans dominate, packed in dense plant populations, a sharp contrast with the wide rows of the 1940s. There is still a large portion of the state's acres planted to hay and pasture, but in modern agriculture, high land values require high returns from cash crops. Finally, an ever-declining number of barns, houses, and outbuildings and the decay of many of them indicate their obsolescence in today's modern agriculture. They are remnants from a significant and fascinating era in which farmers made a host of agricultural, technological, and environmental decisions. Farmers who worked the land during the postwar period not only accepted the industrial ideal but made it their own.[15]

Part One CHEMICALS

One

Insecticide *Time for Action*

In 1943 a new and threatening crop pest, the European corn borer, was in the Iowa cornfields. Iowa farm journalists sounded the alarm about potential crop damage; one headline read "Borer Racing across Iowa." European corn borers had crossed the Mississippi River from Illinois the previous year without attracting much attention, although the species had plagued farmers in the eastern Corn Belt for years. First introduced into the United States in the early 1900s, most likely as stowaways in broomcorn imported from either Hungary or Italy into Massachusetts, European corn borers migrated into Ohio in the 1920s and then across Indiana and into Illinois during the 1930s. Borer moths lay eggs on corn leaves in May, and when the worms hatch they begin to eat the leaves before boring into the stalks where they transform into pupae. In August those pupae emerge as moths and lay their eggs on corn leaves, but this second brood inhabits the cornstalks all winter and emerges the following May to repeat the cycle. Depending on the degree of infestation, these insects prevent ears from developing on the corn plant or even cause them to drop to the ground before harvest. Sometimes the stalk damage is so severe, the developing corn plants blow down in strong winds. By mid-1943 corn borers had spread all the way into the central counties of Iowa and were causing significant losses in the corn crop.[1]

During the war years, extension experts and farm journalists prescribed the same cultural control treatments for corn borers as they invariably prescribed for almost all insect pests: chopping cornstalks after harvest, plowing under all remaining stalk material, making fodder or silage out of green corn, and delaying the planting so as to avoid the worst part of the infestation. All of these tactics interdicted the life cycle of the insect, denying it habitat and thereby reducing the threat for the next year. But in 1946 Iowa farmers began using a new tool to kill corn borers as well as other insects that attacked livestock and plants. Dichloro-diphenyl-trichloroethane (commonly known as DDT) had been developed as an insecticide by the Geigy Company in Switzerland in 1939 and used effectively and with great publicity during World War II. The U.S. Army used DDT to check a typhus epidemic in Naples in 1943, as it killed the lice that carried the typhus. In 1945 Iowa State extension staff worked with county agents and farmers across the state to demonstrate the efficacy of DDT on agricultural pests and thus inaugurated a new era of chemical insect control in Iowa agriculture.

Between 1945 and the early 1970s, farmers used insecticides to supplement and even replace cultural techniques to reduce insect populations. Farmers experienced varying degrees of success in manipulating their environment, however. By 1950 farmers were facing two challenges to their independent chemical strategy. The first was a practical challenge to their daily operations. They discovered that some pests developed resistance to insecticide or naturally tolerated it. Insects that survived treatment could produce offspring that were resistant to the chemicals. The second problem was one with effects far from the farm. Government researchers reported that residues of insecticides could be found in dairy and meat products from treated animals, which posed a threat of unknown severity to consumers. Farmers turned to new chemicals to solve these problems while facing federal regulations to avoid passing poisons on through the food chain. Ultimately farmers relied on chemical control as their first line of defense for some pests but not for others, using chemicals when this would help them maximize their investment. As the 1970s began, farmers could no longer use DDT and some of the chlorinated hydrocarbon insecticides they had relied on, since the U.S. Department of Agriculture (USDA), U.S. Food and Drug Administration (FDA), and state of Iowa banned them due to health concerns. Instead the farmers had a new array of chemicals at their disposal.

Farmers used insecticide to meet their needs, accepting parts of the chemical control message and rejecting others. Chemical insecticide had become a fixture of industrial agriculture before 1972, but there were important distinctions between the types of infestation, types of chemicals used, and even the types of farmers who used insecticide. While Iowa farmers treated crops and livestock for dozens of insect pests in the years after World War II, the focus in this chapter is on three of the most widespread and commonly treated pests: the European corn borer, several fly species, and corn rootworms. DDT treatment for corn borers peaked around 1950 before it faded during the 1950s, giving way to the use of corn-borer-resistant hybrid seed. In the case of fly control, an increasing number of farmers used DDT and other chemicals during the postwar period. By 1972 livestock producers, especially dairy farmers, universally used chemical fly control. Insecticide treatment for soil insects such as rootworms increased dramatically over the course of the 1950s and 1960s, outpacing the use of chemicals for corn borer control. Iowa farmers frequently defied the directions from manufacturers and experts on how to use chemicals, responding to some parts of the chemical-control message and not others. Chemical manufacturers advised farmers, "It's time for action!" when "pests attack your pocketbook," but farmers acted in their own ways and in their own time.[2]

There were few options to stop the European corn borers that were posing a threat to farmers' crops at the height of World War II. Some farmers took the advice of extension staff and farm journalists to plow under cornstalks and use other control techniques to break the borers' life cycle, but they did not stop the spread of these insects. Estimates of crop losses due to corn borer infestations in the 1943 crop year were as high as 4 percent of the corn crop from Indiana to eastern Iowa. Extension staff members characterized the corn borer as "the biggest [insect] problem facing Iowa farmers" in 1945, noting that estimated yield losses that year amounted to six million dollars contrasted with just two and a half million in 1944. By 1948 corn borers had been recorded in every county in the state.[3]

The European Corn Borer and the Rapid Rise and Fall of DDT

Farmers experimented with DDT for corn borer control in 1946 at the urging of Iowa State extension entomologist Harold "Tiny" Gunderson. Experiments with sweet-corn growers showed 90–100 percent control of corn borers in treated fields. Farmers read about a test at Kankakee, Illinois, where 85 percent of the pests were destroyed by dusting plants, resulting in yield increases of up to 25 percent. A writer for *Wallaces Farmer* noted that "It may be some time before the average farmer can use this method of combating borers," since the cost, estimated at $15 per acre, was much more than farmers could afford. It was not long before chemicals for corn borer control became an affordable reality, however, as extension personnel and advertisers promoted the economics of killing corn borers.[4]

Outreach took several forms. Pest-control meetings, speaking engagements, radio spots, and later television programs were important techniques for extension professionals to inform farmers about corn borers and control techniques. In 1945 extension entomologists drafted a script for a talk on corn borer control for use at 4-H club meetings. They stressed cultural control techniques: deep, clean plowing of cornstalks before May 15, planting late, and using hybrids with strong stalks and ear shanks. In 1946 the state entomologist and the agricultural engineers collaborated to sponsor seventeen demonstrations of clean plowing in eight counties. In March 1947 Gunderson distributed to each county extension director a humorous skit written by extension drama specialist Pearl Converse for use at township-level meetings. The skit highlighted both cultural and chemical control methods.[5]

Extension experts were consistent in their advice to balance cultural and chemical control techniques, but DDT was the most appealing part of the pest-control message. Farmers listened to the clean-plowing talks and attended the demonstrations to learn what they could do to stop the borers, but the state entomologist concluded that "DDT appears to be the best single weapon now available against the corn borer." He cited instances of farmers who used DDT to gain higher yields in fields that were free of corn borers. A cooperative venture with the Pioneer Hi-Bred Company on a farm in Durant yielded from twenty to twenty-eight bushels more corn per acre in DDT-treated fields than in untreated fields. These were significant gains when the average yield per acre was approximately fifty-five bushels. Without denying the value of cultural techniques, Gunderson and others made no such claims for their effectiveness, which indirectly encouraged farmers to rely on chemicals.[6]

In 1948 state extension leaders mounted a determined, sustained campaign to promote DDT for corn borer control. Gunderson articulated four points in the borer-control program, which included cultural techniques such as purchasing resistant hybrids, early planting, and clean plowing, but as in 1947 his main emphasis was now on DDT. He stated that every dollar spent on DDT would return from five to ten dollars to the farmer through yield gains. The writer of an article titled "Use DDT to Kill Corn Borers" included information on cultural techniques, but reflecting Gunderson's optimism about DDT, he also emphasized that chemical treatments would make delayed planting unnecessary. The editors of *Wallaces Farmer* claimed that "Old Ways Failed to Stop Corn Borers" and advised farmers to "use every method" to control them, but then they systematically showed the weaknesses of each cultural technique and concluded that spraying with DDT was "about the only corn borer control method that the individual farmer can be fairly sure of." In November a USDA and Iowa State survey reported that a record number of corn borers would be wintering in Iowa. Observers predicted that more farmers would use DDT in 1949, since farmers and entomologists pinned "most of their hopes in the corn borer battle on chemical treatment."[7]

A blanket application of DDT to cornfields was the quick and easy answer for infestations of first- and second-brood borers. Suggested treatment per acre in 1950 was one and a half pounds of DDT per acre, per treatment applied as dust or spray depending on the equipment at hand. The important application lesson was to get the chemical on the leaves and in the whorl of the plant (the place where the leaf emerged from the stalk) since this was where the borer larvae penetrated the stalk. Many farmers who wanted to do their own application made or purchased sprayers in the late 1940s in order to spray insecticide and herbicide. Custom applicators were available in the late 1940s, but a sizable minority of farmers believed it was a good idea to do their own work, as indicated by the remarkable increase in sprayers on Iowa farms, from five thousand in 1947 to almost forty-two thousand in 1950.[8]

Farmers who tried DDT were generally pleased with it, especially when they had a chance to see side-by-side comparisons of treated and untreated corn. John Ostercamp of Hancock County followed the advice of his county extension director in 1949 and performed a double test, leaving unsprayed check strips in the corn he treated and a few sprayed rows in the corn he did not treat. He found that in his unsprayed sections the corn yielded seventy-eight bushels per acre, while his sprayed sections yielded over eighty-five bushels per acre. A Kossuth County farmer obtained as much as fifteen bushels more corn to the acre on his sprayed acres. Farmers complained about unskilled or crooked fly-by-night custom sprayers who promised to spray DDT but either applied improperly mixed chemicals or substituted some other spray, using "the magic name of DDT to fleece farmers." Isolated complaints of fraud, however, did not discourage many Iowa farmers from using DDT.[9]

Chemical treatment for European corn borers had a rapid rise. In 1947 farmers treated only about 50,000 acres of Iowa corn with DDT. The next year they treated approximately 167,000 acres. In 1950 and 1951 farmers treated almost 1.735 million acres, representing 15 percent of the 11 million acres of corn in the state. This increase was due to the hype surrounding DDT, the severity of the corn borer infestation, the ability of the extension system to get the message out about the threat of corn borers, and farmers' willingness to innovate. The state entomologist had mobilized 4-H boys to conduct corn borer counts and had reported damage caused by the borers from across the state for the first time in 1944. These estimates allowed entomologists to predict the extent of infestations for both broods of borers. In theory, statewide estimates permitted farmers to know either if they were at risk for midseason second-brood borers or if they needed to plan for a heavy infestation from the first brood in the following year.[10]

The use of DDT for corn borer control declined just as rapidly as it arose. By the 1960s farmers were treating only a small number of acres with occasional spikes, depending on the degree of infestation. A significant reason for the declining number of DDT corn borer treatments was the relative difficulty in actually observing the insects. Corn borer egg masses on the leaves are no larger than a quarter of an inch long, while the borers themselves are thinner than a pencil lead and approximately one-sixteenth of an inch in length. Farmers who succeeded in finding corn borers then confronted the question of determining the extent of the infestation. Extension experts argued that the population had to cross a specified threshold of fifty egg masses per hundred plants to make treatment cost effective. To discover an infestation, farmers conducted a random sample of a field. They walked into the field a specified distance, counted egg masses on one plant, turned ninety degrees, walked another specified distance, and then counted egg masses again.[11]

Even if farmers were aware of corn borers in their fields and performed the egg-mass counts, they did not necessarily perceive the borers as posing a major threat. In early 1946 a panel of farmers at a corn borer conference in Cedar Rapids reported that farmers would not adopt control measures "until after their pocketbooks have been hurt to the extent of 10 to 15 bushels of corn per acre," implying that farmers could tolerate losses of 15–20 percent of the crop yield. Similarly, the Benton County extension director confirmed, "The farmer who only has a moderate infestation does not do too much about them." As late as 1964 T. A. Brindley, the director of the USDA's Regional Corn Borer Laboratory in Ankeny, Iowa, echoed this concern, adding that years of good growing weather compounded the problem. "It's difficult for a farmer who was growing 50-bushel corn 5 years ago and is growing 100-plus bushels now to be concerned about borers," he observed. Treating for corn borers also coincided with haying season, which made treating for borers a "nuisance" to farmers who wanted to make hay while the sun was shining.[12]

A comparison with chemical weed control helps make sense of the farmers' decision not to spray for corn borers. Farmers with fields infested by weeds could easily see the extent to which weeds were taking over a field even before the weeds were out of control, making it easy to decide to use herbicide. Iowa farmers had contended with weeds for generations (unlike the European corn borer), which made the transition to herbicide much easier than the transition to insecticide.

The most important reason for the decline in treating for European corn borers was the development of new hybrids that either tolerated the insects or were resistant to them. Extension staff members discussed biological control in the mid-1940s and suggested farmers select hybrids with strong shanks and stalks that could withstand some degree of infestation. In the 1950s, seed companies responded by breeding hybrids specially suited to withstanding the borers. Researchers collected borers and egg masses to place in the whorls of the growing parent corn plants. Corn that survived the infestation in good condition became the parent stock of new hybrids. The inbreeding and development of the crosses needed to produce enough hybrid seed to put on the market took time, but each season seed-corn companies offered new and more resistant hybrids. Dr. Brindley of the Regional Corn Borer Laboratory argued that the development of "resistant" or "tolerant" hybrids was one of the most important reasons that farmers by the 1960s were paying less attention to chemical treatment.[13]

Selecting hybrids that could withstand borer infestation was the economical way to fight the insects. Iowa farmers were already committed to buying hybrid seed, so there was no new investment needed. Using resistant hybrids as a preventative approach was potentially more cost effective than chemical rescue treatment. In addition to the additional costs, spraying also conflicted with other pressing tasks such as haying and forced farmers to neglect other profitable work. Furthermore, by the 1950s farmers were facing restrictions on feeding DDT-treated silage to beef and dairy animals, since ingested DDT was stored in the fat of beef and dairy products. Charles Havran of Benton County never sprayed for corn borers. His son recalled, "The big thing . . . was buying hybrids that had stronger stalks." Robert Nymand of Audubon County echoed this, noting that he did not treat his corn for borers in the 1950s and 1960s.[14]

Farmers who treated for borers did not necessarily make it a regular practice. Farm record books show that treatment for corn borers was sporadic, not only inconsistent from year to year but also seldom used on people's entire acreage. Rudolf Schipull of Wright County sprayed for borers in 1951 and 1954. The first year he treated only twenty-nine acres with aerial application at a cost of $50.75. In 1954, two years after he retired from active farming, he spent $186.00 on DDT. He neglected to mention how many acres he treated or the method of application. Joseph Ludwig of Winneshiek County purchased 117 gallons of DDT for corn borer treatment in 1949, and this is the only mention he ever made of purchasing DDT for his crops. William Adams of Fayette County purchased DDT just once before his record keeping ceased in 1959. He bought eighty-three gallons of DDT in 1950 without indicating how he planned to use it. The experiences of Schipull, Ludwig, and Adams confirm that using chemicals to treat for corn borers was the exception rather than the rule.[15]

Results from a 1959 survey of one hundred Adair County farmers show the relatively low degree of effort given to chemical control of corn borers after the initial excitement of the late 1940s. Ninety-five farmers participated in the survey (approximately 5% of the farmers in the county), and seventy-nine farmers reported that they never treated for corn borers. One person who treated in the 1958 crop year noted that he did so to obtain higher yields, but most farmers explained why they did not use DDT. The most common response was that treating for borers did not "do any good." Others reported that the infestation was not bad enough, they lacked time to do it, or they believed it would not solve the problem unless their neighbors treated their fields, too.[16]

Flies and the Triumph of Chemicals?

While the use of insecticide for corn borers decreased in the 1950s, farmers increased their use of DDT and other chemicals for fly control. As early as 1947 experiments indicated that beef cattle gained weight faster and dairy cows produced more milk if they were not bothered by flies. Each blood meal a fly extracted from its host meant less blood supply for the host. Animals twitch, swish their tails, and swing their heads to get flies off, all of which require calories that could be used for production. Unlike corn borer infestations, flies were a visible problem, since dairy farmers were in close proximity to their animals at least twice a day at milking time, and beef producers observed cattle at feeding times. Cooperative extension demonstration projects in 1947 showed that dairy and beef cattle treated with DDT produced an average of over three extra pounds of milk production per day and gained a half pound of body weight more per day than the untreated animals. Many farmers did not keep as detailed records as those who cooperated with extension service to conduct experiments, but they did "watch the milk pail and the cream check." They noticed that, after using DDT, production increased in the summer when fly attacks normally lowered production. Advertisers claimed that for only a few cents farmers could add up to fifty pounds more beef per animal and up to 20 percent more milk with "Pestroy 25% DDT" manufactured by the Sherwin-Williams Company. According to one writer for *Wallaces Farmer*, "The increased gains or extra milk from fly-free cattle are so valuable you can't afford to let flies get started even for a week." Economic concerns made fly-control chemicals an obvious choice for profit-minded farmers, just as planting improved hybrid seed made more sense than spraying for corn borers.[17]

Farmers reported the near miraculous fly-killing power of DDT in the mid- to late 1940s. Bert Brown, a Polk County farmer who raised bulls for artificial insemination, was one of the early experimenters with DDT. In 1945 a group of farmers visited his operation, along with a journalist who observed that, after spraying, there were still some flies in his barn and on the bulls, "but the animals were not being tormented on a hot August afternoon." The next year a dairy farmer from Mitchell County noted, "It works practically 100 per cent in the barn," while a farmer from Dubuque County reported "very good results from DDT application, both on the buildings and on the cattle." A Story County dairy farmer, F. H. Lodgsdon, claimed that "DDT is one of the greatest discoveries for farmers," since in

1946 his farm was free of flies all summer long. A farmer from Greene County declared that "a fellow's foolish if he doesn't take advantage of DDT to protect his livestock from flies." The relative ease of spraying was also impressive. Farmers who sprayed could simply stand on a fence board of a paddock or corral and use a low-pressure sprayer to douse the animals, in contrast to the older technique of driving cattle through a dip tank, which was both wasteful and messy and also panicked the animals.[18]

Inspired by the success of 1946, civic leaders and extension professionals campaigned to rid the entire state of flies in 1947 and 1948. Chemical control was an essential part of this campaign, and extension staff made an intense effort to convince urban and rural Iowans to use DDT. In towns and cities, community leaders coordinated efforts to spray alleys and dumps, even providing residents and small-business owners with free or low-cost DDT to treat window and door screens, kitchen walls, and food preparation areas to eliminate the flies. They encouraged farmers to spray their manure piles every two or three weeks to kill the emerging generation of flies. Farm sanitation was an important part of the effort, too. Gunderson likened using chemicals alone to being a fighter with one hand tied behind his back. Extension staff and journalists goaded farmers to remove manure piles every few days, to clean barns and spread manure onto the fields. The campaign generated interest during the first two years, with as many as 85–90 percent of Iowa farmers practicing a complete fly-control program. It became apparent after two years of the war on flies, however, that it was impossible to eradicate a species that had a life cycle of only several weeks. In 1949 the percentage of farmers practicing fly control declined to 50 percent, which indicates that there were problems with fly-control chemicals.[19]

The widespread use of chlorinated hydrocarbons to control flies created problems that few, if any, of the farmers had anticipated. One of the first problems was that flies developed resistance to DDT. While noting that Iowa farms could be made "fly-free" in 1949, a writer from *Wallaces Farmer* asked "about flies being resistant to DDT?" Extension staff confirmed that flies surviving a treatment could then reproduce and have offspring that would be resistant to chemical attacks. Farmers who had used only DDT for the previous two years needed to try other chemicals such as chlordane, benzene hexachloride, or toxaphene. In the early 1950s the discussion of chemical-resistant flies and the new chemicals available to treat them was a regular feature in fly-control articles in *Wallaces Farmer*.[20]

Another problem with the use of DDT appeared in the health risks to animals and humans. As early as 1946 state extension experts were cautioning that there were hazards associated with DDT, a fact routinely mentioned by the farm press. Gunderson warned farmers that DDT, like any poison, was harmful to people; oil-based sprays should not be used on livestock; and farmers needed to follow manufacturers' directions.

Oil-based sprays were more readily absorbed through the skin or ducts, which meant that the chemical was getting to the animal, not the insects it was meant to kill. In 1953 extension pest-control experts stated that DDT was no longer approved for application on dairy cattle or in dairy buildings. By 1960 there was a "general crackdown" on the misuse of farm chemicals by FDA representatives, who examined animal-tissue samples and products specifically for insecticide residues from chlorinated hydrocarbons such as DDT, dieldrin, aldrin, chlordane, toxaphene, heptachlor, endrin, and methoxychlor. FDA representatives would track serious insect infestations, go to that area to find out what insecticides were being used, and conduct tests on animal products from that area to determine if there was any evidence of chemical residue in excess of the federal limits. They promised to confiscate butter from creameries where they found any trace of DDT.[21]

Sanitation received renewed emphasis in fly-control programs after new guidelines were issued from the USDA and FDA restricting the use of DDT on dairy animals. In 1949 *Wallaces Farmer* noted that, while most farmers were familiar with DDT, it should never be used on dairy cows, in dairy barns, or in rooms for separating cream or storing milk. Most fly-control articles began with a statement stressing the importance of sanitation as the first step in fly control. "Chemicals won't do the job alone," one writer opined. "You have to get rid of fly breeding places. That can mean some extra work." Farmers who failed to clean out barn lots, cattle sheds, and hog pens could not expect any significant success from the use of chemicals. Manure piles growing outside the barn door all winter should be hauled to the field, since these piles were places where flies reproduced. In response to a letter from a Bremer County farmer about chemical control, the editor noted the farmer did not mention that "the best fly control measure" was "sanitation—[the] destruction of places where flies can breed." While chemicals were an important part of the program, sanitation had to be the first step to prevent fly populations from rebounding.[22]

The sanitation message was slow to take hold, however, since farmers had such high expectations of the chemicals. In 1952 Val Racek of Story County had trouble getting rid of flies around his hog house, even though he had used DDT, chlordane, lindane, and other chlorinated hydrocarbon chemicals to kill them. On the suggestion of Earl Raun, an extension entomologist at Iowa State, Racek began to remove manure every three or four days, which brought a significant reduction in the fly population. Racek's impulse to rely on chemicals to control flies made sense in the context of the heavy attention that experts, journalists, and farmers had given to the killing power of chemicals. Sanitation practices required more work than chemical control, a fact that made chemical control appealing and made the sanitation message so difficult to sell to farmers who had experienced success with chemicals.[23]

For all these setbacks for chemical fly control, it was much more popular than chemical corn borer control, especially for farmers who specialized in livestock production. The extent to which dairy farmers had begun to rely on chemicals can be seen in a February 1951 *Wallaces Farmer* survey of dairy breed association members. Over 90 percent of the respondents reported that they believed it was profitable to use fly spray on dairy cows and barns, even though only three out of four were satisfied with the particular products they used. Most farmers used DDT to spray their cow barns, and 21 percent used DDT to spray their cows, with 32 percent using methoxychlor and 27 percent using lindane, even though neither DDT nor lindane were cleared for use directly on dairy cows. When asked "Did you ever hear that methoxychlor was the only residual (long-lasting) fly spray that was safe to spray on dairy cows?" only 43 percent of the farmers surveyed responded yes. This suggests that many farmers were unaware of the guidelines for chemical use. It is also possible that farmers failed to appreciate the risks of chemical use. In 1954 a farmer from Taylor County wrote to the editor of *Wallaces Farmer* stating that he had purchased some lindane concentrate that was labeled not for use on cattle. "But flies were bad," he stated, "and I sprayed the cows anyway," asking what harm the spray would do to the cows. This attitude of "shoot first and ask questions later" indicates that some farmers wanted the quick and easy way to relieve animals plagued by pests and to maximize farm profits.[24]

New products were on the market by the early 1950s to provide a degree of balance in chemical fly-control programs and thereby avoid resistance and residue problems. Methoxychlor was approved for direct application on dairy animals until the mid-1950s, when it, too, was restricted for use only on beef cattle. By 1959 methoxychlor was approved only for use on the walls and ceilings of livestock shelters. In 1955 manufacturers introduced malathion and diazinon, organophosphate insecticides that experts touted as safe and easy-to-use chemicals that killed DDT-resistant flies. Malathion and diazinon could be used in places DDT could not, such as dairy barns. Organophosphates were not, however, approved for use on dairy cows. Regulations permitted farmers to use methoxychlor on dairy and beef cattle, although they could still use DDT or malathion on pastured beef cattle. By the late 1950s fly repellents such as pyrethrin were the only chemicals that could be used on dairy cattle. Gunderson acknowledged that these repellents lasted only about twenty-four hours, but he also stated that "there isn't anything available now that will do the job safely and last longer under Iowa conditions." These chemicals gave farmers more flexibility in controlling flies and allowed them to continue chemical fly control, which was easier than sanitation and solved the problem of resistance.[25]

Throughout the 1950s the farmers' use of fly-control chemicals steadily gained momentum after the initial surge of the late 1940s. In 1952 extension entomologists reported that Iowa farmers treated approximately 45 percent of dairy cattle with insecticide for flies. That number increased steadily throughout the decade, with 50 percent in 1954 and 94 percent by 1960, surpassing the previous high levels of 1948. Similarly, the percentage of farmers using insecticide on beef cattle increased from 7 percent in 1952 to 61 percent in 1960. Farmers with beef herds were aided by the development of a back-rubber-type applicator that farmers could build, which allowed cattle to treat themselves. The early messages from experts about the potential of DDT and their own experiences using it to kill flies made a big impression on many farmers.[26]

Farm records show the high level of interest in chemical fly control. Rudolf Schipull purchased a variety of fly-control chemicals in the 1940s and 1950s. He bought his first DDT for spraying cattle and buildings in 1948, an unidentified fly spray in 1950, and a few dollars' worth of lindane in 1953 and 1954. He bought other fly-control products in small quantities in 1955 and spent $29.41 on five gallons of toxaphene and other insecticides in 1958. Joseph Ludwig bought one dollar's worth of DDT in 1947 and purchased a product called Dustone for cattle in 1954, 1955, and 1956. William Adams bought some DDT in 1948 and eighty-three gallons of DDT in 1950, although it is unclear how he intended to use it. It is likely he intended it for fly control, since he made his DDT purchases on June 23 in 1948 and on June 24 in 1950, which matches the dates he bought other kinds of fly spray in 1951 and 1957. Schipull and Ludwig bought fly-control chemicals in late June, though they also made purchases at other times of the year. By the end of the 1950s these three farmers had all experimented with DDT before moving on to other chemicals.[27]

Farmers hesitated to give up their tried and proven tactics, especially DDT and other chlorinated hydrocarbon insecticides, even after more than a decade of warnings about the potential risks. Farm records show that some farmers modified their chemical use as regulations changed, but many farmers did not. In 1960 Gunderson wrote an impassioned—but confidential—appeal to county extension directors to redouble their efforts to educate farmers about the proper uses of fly-control chemicals. Gunderson was appalled after he examined a survey of Iowa farmers' chemical use, which showed that 64 percent of Iowa farmers used restricted residual chemicals on dairy cows, including DDT and lindane, chemicals that had not been cleared for use on dairy cattle since the late 1940s. "The worst shock," Gunderson noted, "came when I read that nearly 24% of Iowa dairy farmers were slapping a chemical on their cows whose ingredients they *didn't know*." Gunderson recognized the gravity of this situation and warned that the FDA could shut down local creameries and dairies that were the sources of poisoned dairy products. The next year the Clinton County extension director reported that many dairy farmers did not know what precautions to take when they sprayed cows and did not know which products were approved for use on dairy cattle. Two farms in Fayette County were quarantined in 1965 after inspectors found dieldrin residues in milk from those farms. Milk samples from each of the thirty-three cows from those farms tested positive for dieldrin contamination. That year thirteen Iowa dairy farmers had to dump their milk, in one case for 150 days, because of residual insecticides in their milk. The campaign to educate farmers about the merits of chlorinated hydrocarbon insecticide and to promote its use had worked too well. Farmers adopted these insecticides and used them liberally to achieve good fly control but, in the process, potentially undermined their own livelihood.[28]

Rootworms and Resistance

Efforts to control soil insects such as the corn rootworm resembled the widespread adoption of insecticide for fly control, although chemical treatments for the soil-insect complex did not get under way until the 1950s. The opening phase of the campaign began in 1951, when farmers, extension entomologists, and experiment station entomologists cooperated to establish demonstration plots in Dubuque, Butler, Wright, and Boone counties. Cooperators tested five chlorinated hydrocarbon soil insecticides, including aldrin, chlordane, dieldrin, heptachlor, and lindane, to show that farmers could control rootworms and gain up to thirteen bushels more corn per acre. The success of these tests encouraged extension staff to set up more demonstration plots in 1952. In the mid-1950s manufacturers sold fertilizers mixed with insecticide that accelerated the rate of adoption, since the combination of fertilizer and insecticide reduced the number of trips across the field. In 1959 insecticide granules—which could be broadcast or applied in rows, making the treatment for soil insects easier than ever—were on the market.[29]

Farmers readily took up soil insecticide for rootworm control. In 1953 twenty-six county extension directors reported that before planting 85,433 corn acres farmers used soil insecticides to control soil insects, primarily corn rootworms. Extension professionals encouraged farmers to practice crop rotation in order to control rootworms, but insecticides were the popular control solution. Every year of the 1950s more farmers used more insecticide to treat for soil pests in their fields, growing from over 1 million acres in 1954 to almost 2.2 million in 1958 and 5.4 million in 1960. In a survey of farmers' attitudes concerning chemicals, published in 1958, 93 percent of farmers indicated that they were satisfied with their soil insecticides. These farmers discovered that treatment for soil insects was relatively easy, since it could be done before planting or at planting time. Farm journalists counseled that effective control of soil insects depended upon the application of insecticide prior to infestations. Farmers liked the idea of using insecticide on second- or third-year cornfields, which were more vulnerable to a buildup of soil insects than fields in the first year of the corn rotation. Two Dallas County farmers learned that their pre-plant broadcast applications in 1953 helped boost their corn yield by ten bushels in comparison to their untreated crop. Rudolf Schipull and Joseph Ludwig purchased starter fertilizer with insecticide in the late 1950s and into the 1960s. Schipull purchased fertilizer with aldrin for the first time in 1959, spending an average of $92 per year in 1959, 1960, and 1961. He spent an average of $254 per year from 1965 to 1967 on both insecticide-fertilizer combinations and soil insecticide, including aldrin and Aldrex. Joseph Ludwig bought $477 worth of fertilizer with aldrin in 1965 and began to use soil insecticide on a regular basis in 1973.[30]

Soil insecticides used with herbicide and chemical fertilizers were among the reasons that farmers abandoned traditional crop rotations of corn, oats, and hay, which Iowa farmers had used in the late nineteenth and early twentieth centuries to keep crop production high, since crop rotation prevented the buildup of soil pathogens and insect and weed populations. Rotation also introduced nitrogen-fixing plants such as clover to help corn plants grow. Chemical farming meant that farmers could now manage this balance through chemical treatments, which allowed them to grow corn year after year in the same ground. Gunderson urged farmers not to plant corn more than two years in the same field, but the very fact of his protest indicates that many farmers had abandoned rotations that included two years of legumes or grasses in every five-year cycle. This shift toward continuous corn cultivation resulted in changes in the insect populations and especially in the buildup of chemical resistances in soil insects.[31]

The resistance problem in the soil-insect complex emerged in 1961 as farmers of necessity reduced populations of northern corn rootworm. That year the western corn rootworm, a species resistant to chlorinated hydrocarbons, entered Iowa from Nebraska. Unlike the northern corn rootworm (Iowa's dominant rootworm species at the mid-twentieth century), the western corn rootworm was resistant to the chlorinated hydrocarbon insecticides such as aldrin, dieldrin, or heptachlor. The western corn rootworm was a minority species in 1962, comprising an estimated 10 percent of the total rootworm population, and so extension experts did not recommend a change in chemical tactics. They urged farmers to use aldrin and heptachlor, applied at different rates depending on whether the farmer treated only the row or broadcast the insecticide across the entire field. For row treatment, experts recommended application of both insecticides at the rate of one-half pound per acre, while the rate for broadcast treatment was two pounds per acre for first-year corn and only one pound for land in continuous corn, presumably because there was some carryover effect from the first year. By 1963, however, the problem had grown from "only four or five fields" in western Iowa to cover one-third of the state. By 1963 even the northern corn rootworm, a pest that had been controlled by aldrin and heptachlor, showed signs of resistance. In 1966 resistant corn rootworms were present in all Iowa counties. Farmers faced a resistance problem much like the one they had confronted in fly control during the 1950s.[32]

A new and more expensive family of insecticides, the organophosphates, was now thought to be the solution. Diazinon and Thimet, two organophosphate chemicals, were the new killers. Farmers learned about the spread of resistant insects in late 1963, when Gunderson announced that the western corn rootworm was present in most of western Iowa, with "severe problems" in the western third of the state. Joe O'Hara of Page County expressed his interest in the new chemicals in 1964. "I'm going to use an organo-phosphate at planting time on all my corn, first year as well as second year," he wrote. Then he added, "I don't think I can afford to take the risk [of not using it]." An advertisement for the American Cyanamid Company featured a scare tactic. Farmer Neal Van Beek of Sioux Center, pictured with a bag of Thimet 10-G, stated, "We can't expect a crop in 1965 unless we rely on THIMET." Gunderson advised that rotation was one of several alternatives for rootworm control, although a 1964 article in *Wallaces Farmer* only included chemical options as realistic choices. Using aldrin and heptachlor was the least expensive chemical choice, with costs of approximately $1.50 per acre. Diazinon and Thimet, the new chemicals, cost approximately $2 or $3 per acre. Farmers could justify this extra expense if they lived in the higher-risk western part of the state or if their chlorinated hydrocarbon treatment failed in 1963.[33]

Safety and Regulation

The new chemicals were more toxic than the chlorinated hydrocarbons and, as a result, required special handling procedures and provoked more government regulation. Farmers understood that DDT and other insecticides were dangerous, but years of experience showed that most farmers were not at immediate risk from incidental exposure to chlorinated hydrocarbons. Only a large dose was deadly in the short term. Risks from prolonged exposure were more severe, though, because chlorinated hydrocarbons were stored in the fat cells and continued to build up over the years. By contrast, exposure to small amounts of organophosphate insecticides could be lethal. They inhibit proper nerve functioning and prevent the transmission of nerve impulses to vital organs.

Beginning in the mid-1960s, insecticide safety became a much more prominent feature in the coverage of farm chemicals because of serious incidents involving animals and people. At the beginning of the 1965 planting season, Gunderson reported that a farmer had been poisoned while he applied organophosphate chemicals in a tailwind. The man's symptoms included constricted pupils, severe internal cramps, excessive respiratory tract secretion, headache, and weakness. From 1967 through 1970 there was one reported case of livestock poisoning every two weeks, although not all of the incidents could be blamed on the farmers. One Iowa farmer accidentally killed thirty-six steers when he purchased and fed them contaminated feed. The worker who had loaded the cobs for the customer accidentally broke some nearby bags of aldrin, which then became mixed in with the cobs. A study of pesticide accidents from the late 1960s indicated that almost three-fourths of accidents resulted from three causes: failure to follow directions, improper storage, and chemicals placed in improperly marked containers. Farm journalists and extension safety experts urged farmers to "Play it safe with rootworm chemicals" and prescribed a long list of safety procedures, which included using respirators, completely covering the skin with clothing and the eyes with goggles, frequent bathing and washing of clothes, and applying chemicals in calm weather.[34]

In spite of attempts by extension staff and farm journalists to educate the public, people were confused about the new safety equipment. Gunderson confessed that even he did not understand what kind of equipment should be used, and he notified extension cooperators of these problems at the beginning of the 1964 planting season. He claimed that "no one seems to understand the difference between an effective *dust* respirator with a plain filter . . . and a true *chemical respirator*" that was necessary for use with organophosphate insecticides. He urged those who purchased dust respirators to "TAKE THEM BACK!" Dust respirators merely captured the insecticide particles in the mask, forcing the wearer to breath vapors continuously, which, he claimed, "may be even more hazardous than no respirator at all," while the chemical respirator deactivated chemical vapors. The costs of this confusion were severe, since farmers, agribusiness, and the university extension system were now under closer scrutiny after the 1962 publication of Rachel Carson's book *Silent Spring*.[35]

The Iowa Pesticide Act of 1964 was the first major fallout of the post–*Silent Spring* years, a period one Iowa State extension expert referred to as the "After Carson" or "AC" era. This legislation, written in consultation with extension entomologists, became law on January 1, 1964. The law included a provision that all pesticides sold in Iowa be registered with the secretary of agriculture and require state licensure for custom applicators. The state pesticide lab conducted testing on pesticides as well as agricultural products to determine residue levels. This pesticide act was relatively lenient, since it did not require any training or regulation for farmers who applied their own chemicals, but it was recognition that some degree of oversight of farm chemicals was necessary, if for no other reason than to quiet public concerns about the impact of pesticide use. For most Iowa farmers it was business as usual, since they were exempt from special training.[36]

Iowa farmers were divided on the issue of licensure in 1971. When asked how they felt about permitting only licensed operators to apply persistent chemicals such as most insecticides, 42 percent indicated they would favor such a measure, 38 percent opposed a licensure requirement, and 20 percent were undecided on the issue. Younger men, those with some college education, and women generally supported more stringent pesticide rules. A Poweshiek County woman wanted "someone other than my husband responsible for putting on pesticides. I worry every time he uses [them]." Some farmers who opposed regulation feared greater government involvement in farming, while some were pragmatic. One older farmer stated, "I'm afraid the bugs and weeds would take over. How do you get licensed applicators out to do the job when it ought to be done?" Many farmers recognized that there were hazards associated with chemical use, but government regulation was a divisive issue.[37]

As farmers considered the possibility of regulation, the issue of insecticide residue in Iowa's ground- and surface water gained attention. Over the course of the 1960s, stream and well monitoring conducted by the Iowa Water Pollution control Commission indicated that pesticides were showing up in Iowa's waters. From 1966 to 1969 several private wells tested positive for the soil insecticides aldrin and dieldrin as well as the herbicides atrazine, 2,4-dichlorophenoxyacetic acid (2,4-D), and treflan. At various times throughout 1968 chlorinated hydrocarbon insecticides such as dieldrin and DDT were present in the Mississippi, Missouri, Raccoon, Cedar, and Iowa rivers. Public attention was focusing on problems with insecticides in Iowa and the rest of the United States, and these were becoming a concern for everyone, not just farmers.[38]

In 1970, just twenty-five years after the most widely used chlorinated hydrocarbon insecticide was first used on fields and livestock in Iowa, both federal and state governments banned the use of DDT in the United States. The USDA cancelled the registration of DDT, which in effect prohibited its use on most crops and products. That year the Iowa state legislature formed the Chemical Technology Review Board, a group dedicated to the coordination of information about agricultural chemicals among all state agencies. The board included members of state agencies, representatives of the chemical industry, and one farmer, but a smaller advisory committee was responsible for making recommendations to the board for approval. Advisory committee members included representatives from state commissions, Iowa State faculty, a doctor of veterinary medicine, a medical doctor, and two ecologists appointed by the presidents of the University of Iowa and Iowa State University. At an organizational meeting held in July 1970, the board considered, among other topics, the role of chlorinated hydrocarbon insecticides in agriculture. Later that year the committee recommended that the state secretary of agriculture develop regulations to ban the use of heptachlor, DDT, dichlorodiphenyldichloroethane

(DDD), as well as lindane vaporizers, throughout the state, except to control pests menacing public health. Federal or state regulation was not a new feature to farm chemical use, but the complete ban on specified chemicals put farmers who continued to use some of them on the wrong side of the law. The experts—who had originally generated the hype about DDT and then urged caution, as farmers used the chemicals in ways they wanted to use them rather than in ways they were prescribed—now had the force of law.[39]

Iowa farmers began to shape the use of chemicals for insect control with the introduction of chlorinated hydrocarbon insecticides. Extension experts, manufacturers, and farm journalists promoted chemical control as a solution to various insect problems on the farm, but farmers themselves gave some of the most valuable testimony about the chemicals' usefulness. For corn borers, farmers determined that mid-season rescue applications were not the best way to use their time or money, especially since they could use hybrid seed in a preventative approach. Still farmers used more chemicals on more pests throughout the period. For fly control, farmers climbed a steep learning curve, with important challenges such as resistance and regulation emerging along the way. They continued to use the chemicals they believed would do the job, ignoring both government restrictions and advice from entomologists, manufacturers, and journalists. In the case of soil insecticides, applying the chemicals either before planting or at the same time as planting was an efficient use of the farmer's time and money. Farmers made decisions about their use of chemicals based on local conditions and their perception of their needs, not necessarily based on the advice and guidelines they received from other sources.

The 1964 Iowa Pesticide Act, which provided for licensure of commercial applicators, required a degree of specialized training and expertise. So, by the early 1970s, those days of individual freedom in the use of chemicals were passing. The extreme and immediate hazard from the new organophosphate chemicals, compared to the chlorinated hydrocarbons, created a different situation. Farmers who exercised routine caution with the chlorinated hydrocarbons could reasonably expect to be around the next year, whereas those who failed to use the specialized respirators and protective clothing while using organophosphates risked dying in their fields. As historian Thomas Dunlap has shown, the outright ban on DDT was based on its residual effects not on its immediate effects. This new long-term approach to risk assessment of chemical use, coupled with the immediate danger of applying chemicals, meant that farmers were likely to face more regulation in the future. Furthermore, there was new pressure from consumers, environmental activists, and government officials for tighter control of chemicals. The ban on DDT and the regulatory mission of the Environmental Protection Agency (EPA) marked an important turning point in the developing relationship between Iowa farmers and chemical technology.[40]

Herbicide versus Weedy the Thief

Father Time: Why you young villain—you young whipper snap-
per. I'll mow you down, so help me—I'll mow you down.

Weedy the Thief: Save your breath, Grampa. I'll fill your fence
rows and your lawns and your parks and your golf courses
with my Weedy Family. Wait and see.

Miss Verda Land: Father Time—do something.

Weedy the Thief: "*Do* something," she says!! *Do* something!! I've
got you just where I want you. We'll rob your soil and kill
your grass and choke your gardens. We'll fill the world with
weeds—weeds victorious!!!! *Nothing* can stop us.

The Hero 2,4-D: (Bounds in) *I* can stop you.

—2,4-D Skit by Pearl E. Converse and E. P. Sylwester,
ca. 1945–1947

The new weed-killing chemical 2,4-D had been on the market
for less than a year when Iowa State extension botanist E. P.
"Dutch" Sylwester collaborated with extension drama specialist
Pearl Converse to write this informational and promotional melo-
drama.[1] The story began as Father Time, old and tired and armed
with a scythe, tried to defend Miss Verda Land, dressed in green

and carrying a basket representing fertility. The enemy, Weedy the Thief, threatened Verda Land until the Hero 2,4-D arrived and destroyed Weedy with a knapsack-type sprayer. The skit concluded with the marriage between Verda Land and the Hero 2,4-D, signaling a union between landscape and chemicals in postwar America and confirming the optimism most Americans shared about the potential of technology to solve problems.

While herbicides were not new, this particular one was different. It was a synthetic hormone, called a growth regulator, that mimicked the weed plant's own hormones that are present in the growing tips of plants both above and below ground. By stimulating growth the herbicide causes the plant to literally grow itself to death. Unlike older herbicides that killed everything they touched, growth regulators are selective. They kill targeted weed species and, when applied properly, did not damage crop plants. Weeds compete with crop plants for moisture, soil nutrients, and sunlight. Farmers who either fell behind in their work or were careless about controlling weeds traditionally harvested smaller crops than their more vigilant neighbors. Farmers across the United States recognized the potential of this kind of weed-control technique.

In the years after 1945, herbicide manufacturers and advertisers promised that farmers who used their products would achieve remarkable weed control. Advertisements suggested the possibility of weed-free farms, an unheard-of notion as long as farmers relied on cultural weed-control techniques. These cultural techniques varied by crop in the Midwest but typically included planting seed that had been cleaned of weed seeds, spring and fall plowing to bury weeds, regular cultivation during the growing season to check the growth of weeds, and mowing pastures and along fence lines to prevent weeds from spreading. Cultural weed control was time-consuming, however. Iowa farmers spent many hours driving tractors and horses to cultivate their corn crops, mowing fence lines and ditches, and even chopping weeds by hand. Most of this labor occurred while they were pressed by other work such as making hay or silage, which put farmers in the difficult position of having to choose which jobs would get done first, or at all, depending on the weather. The product offered by chemical manufacturers would relieve farmers of many tedious hours spent in cultural weed-control techniques.[2]

But it was farmers, not manufacturers, who gave herbicide its prominence. In the postwar years in Iowa, farmers enticed by the prospect of weed-free farms eagerly purchased growth-regulator herbicide. Rather than passively accepting manufacturers' specifications and extension service guidelines, farmers determined how much to apply and how often to apply it. They often favored herbicides at the expense of cultural techniques. For all the success they experienced in using growth-regulator herbicides to suppress weed populations, there were unanticipated consequences, however. The miracle product 2,4-D did not control all weeds, having no effect on, most notably, grassy weeds such as giant foxtail and quack grass. As Iowa farmers controlled the broadleaf weeds, then grassy species proliferated, which compelled the farmers to use new and more expensive herbicides, sometimes even mixing them or applying herbicide in the soil rather than to the plant. By 1970 farmers were using a new, more complicated and expensive weed-control strategy, and their farms were still not free from weeds. The mixed performance record of growth-regulator herbicides did not compel farmers to lose faith in their chemical strategy, however. Instead, they invested more money in herbicides, using them in record quantities and on record acreages in the annual struggle to keep ahead of weeds.

The Battle against Weeds

Farmers first heard of 2,4-D's promise as an herbicide in 1945 when three chemical companies—Dow Chemical, Sherwin-Williams, and the American Chemical Paint Company—began to sell it for agricultural use. Growth regulators had been used in the 1930s to stimulate growth and promote uniform ripening of some fruits, but their use in deliberate lethal doses did not come until 1941 under the direction of E. J. Kraus of the University of Chicago. The U.S. government experimented with 2,4-D as a tool for biological warfare throughout World War II. Although the war ended before 2,4-D could be used for military purposes, promoters believed it could play an important part in agriculture.[3]

The chemical companies' advocacy of 2,4-D came just at the time when farmers, extension professionals, journalists, and county supervisors were arguing that the weed problem in Iowa was much worse than it had been before the war. Labor-intensive work was required to control the weeds that cut yields, were toxic to livestock, and cost farmers money. Citing wartime labor shortages and the burden of increased production goals, county officials and farm journalists highlighted how "Weeds Won in War Years," as *Wallaces Farmer* put it in early 1947. County extension directors reported on the deterioration of Iowa farms. While farmers attacked weeds in the fields, they were lax in controlling them in the fencerows and ditches, which then allowed the unwanted plants to propagate and proliferate. The Hamilton County director reported that because of manpower shortages "we have many farms now that are fairly well covered with noxious weeds that were fairly clean a few years ago." Canada thistle was a particular problem in Humboldt County, gaining a presence on as much as 95 percent of farms there. When Bert and Vesta Sams moved from Clarke County to Marshall County in 1946, they purchased a weed-choked farm that had been neglected for years. Clifford Sams, their son, was twenty-five at the time and had just returned from military service.

He recalled, "This place was so infested with weeds of every kind; bur-dock, sourdock, Canada thistle. It really took a lot of serious work to get control of it."[4]

Not all farmers were as aggressive about weed control as Bert Sams, however, and in 1947 the state legislature passed a stringent new noxious-weed law. Legislators hoped the new law would compel farmers to control the neglected weeds in their fields, fencerows, and roadside ditches and thereby save the state's farmers millions of dollars lost because of weed in-festations. Declaring war on weeds, lawmakers allowed county commis-sioners to levy taxes to pay for county weed-control employees as well as equipment and materials. County commissioners could notify farm own-ers or tenants of a weed problem and give them five days to cut or spray the noxious weeds. If farmers did not address the problem, the county "weed man" could come out and do it, charging owners for costs plus a penalty fee of 25 percent of costs. Farmers with property adjacent to county roads had to cut ordinary weeds within thirty days of notification or suffer the same penalty. Over a year later, farmers reported that the weed law helped reduce weed populations, but all the farmers surveyed in a *Wallaces Farmer* poll noted that lax or spotty enforcement compromised the law's effectiveness. Some commissioners reported progress in the fight against weeds in 1949, claiming that education and gentle pressure were effective tools in obtaining cooperation.[5]

In this battle against weeds, state and county extension staff members were active in spreading the news about weed-control techniques and pro-moting 2,4-D as well as cultural weed-control practices. Throughout the winter months extension botanist Dutch Sylwester and extension ento-mologist Harold Gunderson conducted weed and pest clinics around the state for groups ranging from a few dozen to over two hundred farmers. From the fall of 1947 through the following autumn, Sylwester conducted 193 meetings with estimated total attendance of approximately 133,000 people. At the 1945 Humboldt County Fair, Sylwester assisted with a spe-cial "weed day" during which farmers could examine a demonstration plot to see the effectiveness of chemical weed killers. During the growing season, Iowa State hosted field tests at different locations around the state. County extension directors also played a vital role in promoting the use of herbicide by setting up demonstration plots, hosting Sylwester for weed meetings, and answering questions from area farmers about herbicide. The Kanawha Experimental Farm in Hancock County was open for touring on July 6, 1948, for farmers to see "what happens to weeds when 2,4-D and other weed killers are used in crops." Implement dealers displayed various types of weed-control tools, reflecting the integrated cultural and chemi-cal weed-control approach favored by extension staff members.[6]

As much as extension staff professionals advocated using chemical her-bicides for weed control, they never recommended these as any exclusive solution but continued to stress the importance of cultural techniques. During the 1940s and 1950s state and county extension staff emphasized the value of chemical control for problem weed patches or for use during wet years when cultivation was difficult, but they never advocated it as a panacea. Extension experts believed that chemicals ought to be the "ace in the hole," not a replacement for cultural techniques. *Wallaces Farmer* writers echoed the experts, noting, "Chemical weed killers are vital to efficient weed control. But they are not a substitute for a number of good farming practices." Sylwester argued that planting clean, weed-free seed for oats and pastures was the foundation of good weed control. Practicing proper rotation and good seedbed preparation and cultivating and mowing weeds in pastures and fence lines were regular admonitions in the farm press and in the talks Sylwester gave across Iowa. Furthermore, experts noted that many plant species such as grasses were resistant to 2,4-D and could be controlled best through cultivation. Throughout the 1960s extension staff and farm journalists continued to recommend cultivation for weed control in row crops such as corn and soybeans. Extension pamphlets on weed control in corn and soybeans began with the standard statements about the primacy of cultural control. Experts and journalists on the farm beat were consistent and insistent in their message of integrated weed control.[7]

Embracing Herbicides for Weed Control

Exciting stories about the success of 2,4-D overshadowed professionals' frequent warnings to balance chemical and cultural practices, however. Reporting on the trend to use 2,4-D on cornfields in 1949, Jim Roe of *Successful Farming* claimed "practically every man who used 2,4-D thinks it put more corn in his crib." A 1949 *Wallaces Farmer* article titled "Bad News for Weeds" detailed the successful experiences of three Warren County farmers who used 2,4-D in cornfields during the 1948 growing season. Bottomland farmers who regularly dealt with wet conditions found that using herbicide allowed them to eliminate weeds that grew rapidly during the times when wet weather delayed cultivation. One farmer reported that some of the Des Moines River bottomland he farmed was so infested with weeds it was impossible to distinguish the rows. "Three or four days later [after spraying], you could see the rows. Those fields ended up fairly free of weeds. Without spraying," he added, "it might not have made any corn." Adolph Erickson of Emmet County explained that in 1948 he used 2,4-D on cockleburs and Canada thistles in his cornfield. In 1949 he sprayed again and claimed, "It really gets 'em. I sprayed just before the corn got too tall to drive thru [sic], and kept the spray off the corn. Another year or two and I'll have them cleaned up." Optimistic testimonials such as these made compelling reading for farmers who understood the costs of getting rained out of cultivation and watching weeds overtake their crops.[8]

Farmers of the late 1940s and 1950s readily accepted chemicals as part of their weed-control system. They had grown up using scythes in their fencerows and had endured stands of flax and oats that were infested with weeds, which cut yields, slowed the harvest, and forced them to cultivate their corn crop three times each summer at the mercy of the weather. These experiences made them eager converts to new chemical techniques. Furthermore, in the years after World War II there were fewer hired men available to do these labor-intensive tasks. In 1949 county directors estimated that approximately 41 percent of Iowa farmers used herbicides. More farmers used herbicides in the northern and western portions of the state where cash-grain and livestock-feeding operations dominated. In the extreme northwestern counties usage was as high as 60.1 percent, while in the extreme southeastern counties it was as low as 26.7 percent. Farmers tended to use herbicide first on fencerows and roadsides, then on pastures, then corn-, oat, wheat, and flax fields. By 1954 more farmers expressed interest in using 2,4-D on cornfields, while in 1963 farmers used herbicide in cornfields as often as they did in fencerows and along field borders.[9]

Even farmers who resisted herbicide use found it could be beneficial. Bert Sams of Marshall County was a firm believer in cultural weed-control practices, but his son Clifford began using herbicide in 1954 after signing a contract to grow seed corn for Pioneer Hi-Bred. The terms of the contract stipulated that Sams use herbicides, which linked him to agribusiness in a new way. After the first year of growing for Pioneer, he justified using herbicide because he made higher profits on seed corn than other farmers made for cash grain. Clifford recalled that his father dropped his opposition to herbicide because of the potential profit that could be gained by spraying. According to Clifford, "We felt if everyone else is doing it and you don't you're going to fall behind." Using herbicide did not match Bert's traditional perceptions of good farming techniques, but he conceded that good farming maximized yields and profits and that businesses such as Pioneer could help him do just that.[10]

The growing number of sprayers in Iowa reflected the growing popularity of herbicides. In 1951 Sylwester and Gunderson claimed that spraying for weed and insect pests had grown from "Babyhood to Manhood in 5 Years." While there were only 85 commercial spray operators in Iowa in 1946, there were 375 the next year, and 800 in 1949. That number dropped to 688 in 1950, but the growth in the number of farmers with sprayers suggests they obtained their own sprayers so rapidly that not all the custom operators could stay in business. Just one year after 2,4-D came on the market, 5,000 Iowa farmers owned sprayers, while in 1948 an estimated 22,407 farmers had their own equipment. In 1949, 30,000 farmers owned sprayers, and in 1950 just under 42,000 farmers could spray their own fields and fencerows. As a 1949 survey indicated, 33 percent of farmers who sprayed owned their own sprayers, but 76 percent owned equipment in 1949, with ground equipment such as tractor-mounted or tractor-drawn sprayers accounting for as much as 90 percent of the total.[11]

Despite the generally enthusiastic adoption of 2,4-D and sprayers, the technologies were not without their difficulties. Farmers' responses to these problems offer powerful evidence for their role in shaping technology. Whether they sought solutions or simply decided to live with the problems, their decisions ultimately defined how herbicide technology would fit into the organization of farming during the postwar years. One fundamental problem was that herbicide did not always control weeds in ways the farmers expected. In 1950 county extension directors reported that 67.3 percent of farmers were satisfied with their spraying and that farmers in most counties would increase the use of herbicides, yet some directors expected a decline in use by as much as 10 percent. In 1954 *Wallaces Farmer* conducted a poll of Iowa farmers to assess how they felt about their chemical weed-control efforts. The Wallace-Homestead pollsters asked: "If you sprayed the following weeds [buttonweed, cocklebur, and smartweed] in 1954, what kind of results did you get?" While only 3 or 4 percent reported "no kill" on the listed weeds, between 57 and 65 percent responded they had "good kill." This left between 32 and 39 percent of farmers who sprayed reporting only "partial kill," a significant minority.[12]

A variety of conditions and circumstances explain the mixed performance record and mixed perceptions of herbicides and the limits of chemical control through the mid-1950s. Extension directors claimed that the most common problem was operator error. Crop injury such as brittleness in corn occurred when applicators used too high a rate or sprayed at the wrong crop-growth stage. Brittleness was a serious problem because farmers who cultivated could break brittle stalks with their equipment. Regular reports of crop injury due to "too much" chemical, "overdose," or "too high concentration" came from across the state as long as extension directors filed special weed reports from 1948 to 1953. A custom applicator from Illinois advised readers of *Successful Farming* that weeds in cornfields needed to be sprayed when the corn was between two inches and eight inches tall, although county directors frequently noted that farmers applied 2,4-D at the wrong time. The speed of the tractor, getting too much spray on the corn and not on the weeds, and spray pressure were also frequent problems that could affect the quality of the weed control or even damage the crop.[13]

Environmental conditions such as temperature, wind, and rainfall also accounted for mediocre herbicide performance. The Hancock County extension director observed that 2,4-D was not very effective on thistles in 1951 because of cold, wet conditions during spraying. An experiment in

Grundy County during the 1957 season illustrates the environmental and human variables that farmers were struggling with in their efforts to achieve weed control. Extension botanist Dutch Sylwester recommended spraying 2,4-D in one sixty-acre field infested with sunflowers before the farmer planted soybeans. Later that year, the Grundy County extension director reported that one-half of the field showed 100 percent weed kill while the other half showed only 40 percent kill. He stated, "This suggests that methods of mixing, temperature of the solution, etc. still has a lot to do with the effectiveness of chemicals in controlling weeds." These considerations could all be present or absent in any given year, making herbicide application a gamble.[14]

Furthermore, farmers did not always know what they were seeing when dealing with this new technology, even when it worked. A Hamilton County farmer used 2,4-D on part of his oat field, only to observe that the oats in the portion of the field he sprayed appeared to be damaged by the spray while the untreated area was green. The farmer contacted his local extension director who brought Sylwester to the farm. When they went into the field they noted that the green they saw from the edge of the field was actually weeds, not oats. Sylwester concluded that the oats were damaged but not by the spray. The oat damage was due to blight, although the county extension director noted that the farmer was reluctant to admit it. In this case the experts concluded that the herbicide worked even though it appeared to have failed, but it took expert interpretation to prevent the farmer from blaming the herbicide for his crop failure.[15]

While the Hamilton County incident was a case of mistaken identity, other crop failures could be attributed to inherent qualities of the chemical and the ways farmers applied it. Farmers with specialty crops found that spray "drift" from herbicide application could cause unintended damage. In 1952 the assistant manager of the Council Bluffs Grape Growers Association notified Sylwester that there was "considerable weed spray damage to our grape vineyards," a fact reflected in the county reports from Pottawattamie County as early as 1950. Growers blamed the railroads. To prevent weeds from spreading across the state along railroad right-of-ways, railroad administrators hired airplanes to spray the weeds growing in ditches along the tracks, but aerial spraying was especially prone to drift. While Sylwester visited Pottawattamie County every year up to 1953, the county director only reported on the problem, with little comment on any proposed solutions aside from urging the railroads to use a less volatile formulation of 2,4-D, which was less prone to drift.[16]

In early 1954 the board of directors of the Council Bluffs association, extension weed-control and horticulture experts, and the county extension director attended a meeting to develop guidelines for minimizing drift. These guidelines included notifying neighbors of grape growers about potential risks, mapping the locations of growers so sprayers working for railroads and power companies to control weeds along right-of-ways and power lines could use cultural practices in areas near vineyards, using a formulation of 2,4-D with larger particles that were less likely to drift, and holding meetings to educate people who spray about the potential for damage from drift. There were other techniques to reduce drift, such as basal spraying rather than applying the chemical to the foliage, but airplane spraying was indiscriminate and, from the user's perspective, much less expensive.[17]

In spite of the 1954 meeting and newly agreed upon procedures, there is no evidence that these efforts resulted in less drift damage. Grain farmers in pursuit of their own advantage were generally unwilling to change their herbicide strategy, despite the conflict over drift damage. Some of those who suffered fought back but often with little success. In 1955 Cecil J. Baxter, a fruit grower from eastern Iowa, expressed concern to Sylwester, informing him that at least five hundred grapevines had been killed by drift and that many other vines had been injured. Baxter blamed the weed specialists, whom he believed had pressured grain farmers to use 2,4-D. In fact extension professionals had always recommended a balance of spray and cultural techniques, not strictly spray solutions. Rather than being overly compliant to the advice of extension specialists, as Baxter claimed, farmers frequently did not heed experts and chose to spray rather than practice cultural weed control. Baxter gained some cooperation from a Lee County employee who agreed not to allow spraying within three miles of Baxter's vineyard and to spray at low pressure with a larger nozzle—to prevent small particles from drifting—and only when the wind was from the opposite direction. Despite this kind of cooperation, the issue remained contentious. In 1964 officials in Muscatine, Lee, Mills, and Harrison counties and the rural portions of Pottawattamie County all banned the use of 2,4-D formulations that were prone to drift. Then the Iowa Fruit Growers Association called for a statewide ban on highly volatile forms of 2,4-D, arguing that damage in 1964 was as bad as in any previous year. Unfortunately for the fruit growers, the wishes of the majority of farmers who raised grain for sale or feed prevailed, and the state legislature, dominated by rural legislators whose constituents favored 2,4-D, did not enact a ban.[18]

In addition to the possible unintended damage it could do to crops, there was also evidence that 2,4-D was more toxic to humans and animals than promoters claimed. Throughout the 1940s and 1950s extension experts and farm journalists urged caution in using herbicides but steadfastly maintained there was little or no risk of injury to livestock or humans. Still they had to remind farmers not to confuse the relatively benign herbicide 2,4-D and the highly toxic insecticide DDT. The farm press reported on tests where scientists fed 2,4-D and 2,4,5-T directly to animals, applied it to feed and forage, and treated their skin with it. Scientists concluded that no permanent injury to livestock ensued. However, tests did show that 2,4-D could cause at least temporary harm to both livestock and humans. In some cases the chemical changed the starches in plants to sugars, which made some plants that were toxic and distasteful to livestock more appealing but no less toxic. One Audubon County farmer recalled that he was listless and fatigued after a day or two of spraying 2,4-D. In 1962 a railroad switchman from Davenport became soaked with 2,4-D after he walked through treated weeds while working in Illinois. He developed a severe rash, and the chemical aggravated his preexisting condition of peripheral neuritis (loss of nerve function in the extremities).[19]

The uneven performance record of growth-regulator herbicide, including evidence of injury to humans and livestock and reports of crop injury, did not overshadow the optimism shared by most farmers, journalists, and extension observers about its potential. Success stories abounded to offset the reports of failures. In general the complaints about growth-regulator herbicides came from minorities such as the grape growers, horticulturists, and an occasional livestock producer. For most farmers the value of herbicide in the field apparently outweighed the potential problems such as damage to farm gardens from spray drift. Farmers found that herbicides offered them at least some degree of improved weed control at reduced cost, which made the limited risk of injury worth the potential payoff in higher crop yields.

Aside from the advantages of saving time in cultivating, gaining the upper hand when bad weather prevented cultivating, and controlling acute infestations, farmers found that chemicals were inexpensive. In June 1948 Joseph Ludwig of Winneshiek County paid $49.22 for five gallons of 2,4-D at a local implement dealership. In July he purchased two more gallons from the same vendor for $19.49. In 1950, however, he patronized the same vendor and paid $4.43 per gallon for eight gallons, with the price varying between $4.00 and $5.50 per gallon over the next few years. William Adams of Fayette spent $8.39 per gallon in 1949, $6.63 in 1952, but only $3.29 per gallon in 1957. He spent an average of $68.00 per year for herbicides, which was an average of 5 percent of annual crop expenses between 1949 and 1958.[20]

Sprayers were also relatively inexpensive, especially since they could be used for insecticides also. In 1949, advertised costs ranged from $219.00 for the Essick tractor-mounted model with a fourteen-foot boom to $230.00 for the Speedy Sprayer with a thirty-foot boom. Tractor-drawn trailer sprayers were even less expensive. In 1958 Kim's Fast-O-Matic trailer model was advertised at $208.95, and the Century trailer sprayer at $186.25. William Adams bought a sprayer in 1949 for $276.11. Charles Havran of Benton County bought a thirty-gallon drum, pump, nozzles, and boom and built a sprayer on his farm. Many farmers chose to make their own equipment, hoping to cut costs by investing their own labor.[21]

Even farmers who purchased sprayers spent very little compared to the cost of other farm equipment. In 1949 when the Iowa State Extension Service published a pamphlet with instructions on how to make a home-made sprayer, the weed-control experts contended, "Tractor units can now be bought about as cheaply as they can be made at home." Joseph Ludwig bought a granular herbicide attachment for his planter in 1961 for $84, while the year before he had purchased a sprayer, possibly a used one, for $70. These items constituted a small part of Ludwig's total investment in machinery and implements. For example, in 1966, the depreciated value of both herbicide tools was $75 out of a machinery inventory of $30,680 at year's end. These expenses were nominal for farmers, making equipment cost a non-issue when deciding whether to use chemical weed killers. The degree of weed control was what really mattered to Iowa farmers.[22]

Shaping Herbicides and Cultivation

The financial advantages of herbicides gave farmers plenty of motivation to adopt 2,4-D, but they did not stop there in their efforts to shape this technology to their needs. They also actively experimented with herbicides in order to achieve even more cost or labor savings, hoping in some cases to reduce even further the need for mechanical weed control. Some farmers and experts began experimenting with preemergent application (applying herbicide after planting but before crops emerged from the ground) in order to get a head start on weed control early in the growing season. They endured numerous failures before they could count on preemergent application as a reliable method of control. Tests in cornfields in 1948 generally gave poor results, although tests at the Iowa State agronomy farm showed that preemergent spraying in soybean fields could be successful. In 1950 and 1951 farm writers were more upbeat about the prospects for preemergent application, but they were careful not to issue a blanket endorsement, having received mixed reports from the field. In 1951 journalist Wally Inman used cautionary language when describing the chances for success of preemergent application: "If you care to spend the money, or are sure the right kind of shower is coming [to ensure absorption], you can do a pre-emergence job to be proud of. Otherwise? It's your money!" Heavy rains could wash away the chemical before it had a chance to interact with the weeds, while dry conditions could prevent the chemical from being absorbed by them. There were too many variables to make preemergent application a reliable practice.[23]

New products on the market promised improved preemergent control during the late 1950s and early 1960s. One Humboldt County farmer planted an experimental plot to show the effect of 2,4-D and two new products —Randox and Sinazon—when applied as preemergents. The extension director concluded that there was some improvement in weed control in the sprayed areas but that the results were "not particularly good." He stated that, where the seedbed had been well prepared with harrowing, "there was just as much or more weed control as where the proper amount of pre-emergence spray had been applied." *Wallaces Farmer* reported excellent results with preemergence weed control in 1960. A 1961 survey indicated that only 3 percent of Iowa farmers used preemergent application but that this number would increase through the decade. Reports on the performance of Randox and another new chemical, Atrazine, were generally positive. Success stories from neighbors and farm journalists encouraged farmers to experiment with this new type of application.[24]

The experiences of farmers from across the state demonstrated this new trend of minimizing cultivation, even before preemergent application was common. In 1951 a farmer from Mills County sprayed corn and followed it with only two cultivations. Two years later several farmers discussed their modified weed-control strategies, emphasizing the importance of herbicide as a labor-saving technology. William A. Fridley of Warren County noted, "I usually cultivate corn three times. But I want to try one or two cultivations this year—that is, if the weather and weeds will let me." An Ida County farmer reported that he substituted spraying his corn for one pass with the cultivator. A custom applicator from Illinois shared his experiences with readers of *Successful Farming*. He believed that cultivation had its place but that a well-timed application of herbicide could save time and expense over cultivating. While herbicides had financial costs, preemergent spray and granules or postemergent spray were cheaper than cultivating. A tractor with a trailer or mounted sprayer used less fuel than a tractor with a cultivator, operating speeds were higher, and the effective width of equipment was greater, thereby saving labor. This degree of flexibility in cultivation would have been unthinkable to farmers before 1945, but it was now possible and attractive to a growing number of farmers during the 1950s.[25]

The shift among farmers to less cultivation was disconcerting to some observers, especially the experts. Extension directors who preached balancing cultural and chemical techniques worried about an overreliance on chemicals. They recalled applications that had gone awry and knew that herbicides were not always effective. In 1953 the Clay County extension director recognized the value of weed killers but feared, "There may also be the other effect of relying to[o] heavily on chemicals and lowering our diligence in other cultural methods of weed control." In 1958 Sylwester, an advocate of an integrated strategy, complained that "too many folks are expecting chemical weed killers to perform the entire control job." The Hardin County extension director observed that some farmers were vigilant about weeds only when they could control them easily by spraying. These caveats were rare, however. Sylwester continued to promote weed-killing chemicals. He probably understood farmers wanted weed control that would save them either money or time that they could then devote to other endeavors, such as specializing in livestock feeding, dairying, or farming more acres.[26]

By 1960 farmers relied on chemicals for weed control, either supplementing cultural practices or replacing one or more cultivations with spraying. But reliance on an almost purely chemical strategy of weed control came with ecological consequences that would ultimately make it more complex and expensive to achieve weed control with herbicides. The very success enjoyed by farmers who used 2,4-D to control broadleaf weeds created an ecological vacuum. As farmers reduced the broadleaf populations, species that previously had coexisted with the broadleafs began to face less competition, and those weeds that were tolerant of 2,4-D began to thrive. As historians of the environment as well as contemporary observers noted, nature responds to human activity in ways that users of technology cannot predict. Just when herbicides were becoming commonplace, observers noticed that a new type of weed was moving into the Corn Belt. A writer for *Wallaces Farmer* opened a 1954 article titled "Grassy Weeds Threaten" with a compelling lead, "Needed: a better method of controlling grassy weeds." Reporting on the discussions at the North Central Weed Control Conference, the author observed, "The threat of grassy weeds to Corn Belt crops is increasing as competition from broadleafed weeds is reduced thru the use of 2,4-D." The following year the magazine's writers echoed the same fact, noting that around 1950 giant foxtail had been present only in southern Iowa but in just a few years had spread to every section of the state. In 1956 the range for giant foxtail was halfway up the state, in Greene County. By 1965 it had become a leading weed in Iowa. Robert Nymand of Audubon County captured the essence of the problem when he stated, "The 2,4-D seemed to feed the grasses." While it was common knowledge that 2,4-D did not control grasses, the rapid advance of grassy-weed species surprised most farmers and threatened to undermine the success of chemical control.[27]

Chemical manufacturers and weed-control experts responded quickly, introducing new types of chemicals, including new growth regulators as well as new contact herbicides that killed everything they touched. *Wallaces Farmer* announced "New Weed Control Chemicals for 1961"—specifically two growth regulators, Amiben for soybeans and Atrazine for corn, as well as a new formulation of the contact herbicide amino triazole, called Amitrol. Experts recommended Randox and Atrazine for giant foxtail control in corn for the 1961 growing season, while soybean growers could use Alanap, Randox, and Amiben. Still 2,4-D continued to be a mainstay for Iowa farmers, since broadleaf weeds remained in fields, pastures, and fencerows.[28]

The new chemicals helped control the grasses, but they were more expensive than 2,4-D and required farmers to spend more money on herbicide in the 1960s than during the 1940s and 1950s. Rudolf Schipull, who began using 2,4-D in 1952, spent an annual average of $11.36 from 1952 to 1958, not including 1957 when he recorded no herbicide or spraying expense. After spending an average of $107.38 per year from 1959 to 1967 (mostly for 2,4-D and custom spraying), Schipull found his herbicide expense changed dramatically the next year. In 1968 he purchased $380.51 worth of Randox, Treflan, Atrazine, crop oil, and Weedout. In 1969 the only herbicide he listed by name was 2,4-D, but he spent almost $300.00 for it, while in 1970 he spent $473.00 for seven different herbicides and custom applications. It is impossible to know the particular weed conditions he was facing, but it is clear that Schipull believed the increasingly expensive chemicals could help control weeds in his fields.[29]

In addition to costing more money, herbicides represented a larger percentage of crop expenses in the 1960s than they had in previous decades. Joseph Ludwig's nominal herbicide expense for 2,4-D and the occasional supplement amounted to an average of 1.5 percent of his total crop expenses from 1946 through 1960. During these years the highest percentage was 7 percent in 1950, and he recorded no herbicide expenses in 1949, 1957, 1958, or 1960. In 1961 Ludwig altered his operation significantly, recording 38 percent of crop expense devoted to herbicide. From 1961 to 1970 his herbicide expenses averaged 29 percent of his total crop expenses. While the average total crop expense for his farm during both periods was similar ($1,698 from 1946 to 1960 and $1,683 from 1961 to 1970), the larger percentage devoted to herbicides shows how these had become a significant part of Ludwig's farm operation.[30]

Farmers like Ludwig who learned how to use herbicide by experimenting with 2,4-D faced an increasingly complicated lineup of herbicides to choose from during the late 1960s. Changing weed populations meant changes in herbicide tactics. As an editor of *Successful Farming* noted in 1969, "So many brands, so many different formulations with different rates of application and different costs, make selection of a weed killer a pretty tough job." With nineteen herbicides for corn and twenty for soybeans available in 1969, the days of using just 2,4-D or perhaps a contact nonselective herbicide such as amino triazole were long gone. The proliferation of new herbicide varieties and names could be confusing, with some chemicals such as Ramrod or Randox appropriate for corn- and soybean fields, but Randox T and Londax cleared only for use in cornfields. A Hardin County extension director remarked in 1960 that farmers wanted to know which chemical to use on a particular type of weed—in contrast with earlier reports that farmers wanted to learn about the handful of chemical treatments on the market. In addition to offering information on the proper use of 2,4-D, the director observed, "Many requests are for information on use and hazards of new weed chemicals," suggesting sensitivity on the part of some farmers regarding the search for the right herbicide for a particular weed problem and their concern about crop damage from misapplication, improper mixing, or carryover. Furthermore, dosages changed over time, as weed problems changed and manufacturers modified product formulas.[31]

The demonstration projects and the specifics of recommended dosages illustrate the complexity involved in selecting one or more herbicides. Henry County farmers saw a variety of chemicals in use in 1966 when a demonstrator near Winfield compared broadcast preemergence application of three chemicals (Atrazine, Ramrod, and Randox-T) with postemergence application of Atrazine combined with mechanical techniques. In 1967 Hamilton County extension staff cooperated with the Stanhope Co-op Elevator and a local farmer to manage a demonstration plot with ten different herbicides on beans and seven herbicides on corn. In 1967 Sylwester recommended nineteen different chemicals that could be used for corn and soybean fields, with each chemical applied at a different rate and at different times. Yields varied widely depending on the chemical used, which all made herbicide use a trickier proposition than it had been during the 1950s.[32]

As herbicides became more complex and expensive to use, manufacturers and experts nevertheless advised that the chemicals were cost-effective. Manufacturers contended that every dollar farmers invested in herbicide gave better returns than a dollar invested in cultivating, which included fuel, labor, and machinery costs. The Geigy Chemical Corporation argued in 1968 that for every dollar's worth of Atrazine, farmers could expect to gain up to four dollars through increased yields. Researchers at experiment stations proved that just a few weeds could decrease yields, confirming the advertisers' claims. One giant foxtail plant per foot of corn row cut yields from 93.5 bushels to the acre to approximately 86.5 bushels. Leland Bentley of Grundy County provided a dramatic example of the yield-increasing potential of the new chemicals with an Atrazine test he conducted in 1965. Bentley applied Atrazine on a cornfield where nut grass practically covered his four-inch-high corn. At harvest time the section where he applied the Atrazine yielded 19 bushels per acre more than the control plot.[33]

It was not only the type of herbicide but the application techniques that changed during the 1960s. Farmers used pre- and postemergent spray herbicides and granules in the 1950s, but the instructions varied from chemical to chemical. During the 1950s most preemergent herbicides could be applied anytime after planting and before the crop emerged, but the newer chemicals required more sensitivity. The Amchem Company noted that there were numerous complaints about their product Amiben, an herbicide marketed for soybean fields. According to a company representative, "it seems to me that the farmer still doesn't fully understand how to use AMIBEN for most consistent results." He further observed that, for best results, farmers needed to apply it at the same time they planted, implying that farmers needed to relearn application practices to match the new chemicals.[34]

Other techniques—such as incorporating spray herbicide into the ground—also required training. Eli Lilly Company's herbicide for soybean fields, Treflan, required mixing it into the top layer of soil to obtain maximum contact with the delicate roots of emerging grassy weeds. The company worked to ensure success for the product, issuing a question-and-answer circular to county extension staff about proper techniques for using Treflan. The circular stressed the need to use a tandem disc and harrow to get the chemical mixed properly with soil. That year the company also published an extensive brochure, "New Dimension in Weed Control," emphasizing "three-dimensional protection against weeds rather than just a thin veneer on the surface." Incorporation was not a blanket approach, since not all herbicides were effective when incorporated and some could actually damage the crop. Chemicals that dissipated in air or decomposed in sunlight, such as Treflan, were good candidates for incorporation. Incorporation, like all the new techniques, required a higher degree of sophistication in the farmer's application. Mixing too deep diluted the chemical. Mixing too deep or too shallow meant the chemical could miss the roots of the weeds.[35]

The growing complexity of herbicides may have been daunting, but it did not reduce farmers' faith in the chemical strategy. They continued to explore new ways to use this technology to their advantage, even when added complexity was the result. With the proliferation of herbicides designed to combat different types of weeds during the 1960s, it did not take long before farmers and researchers combined herbicides to achieve broad-spectrum weed control. As one writer noted in 1966, "Combinations offer the possibility that someday, by mixing two, or perhaps several chemicals, you can tailor a prescription herbicide to your specific weed problem." Farmers mixed herbicides not only to obtain broad-spectrum weed control but also to match local soil or weather conditions, reduce the risk of residue of a full-strength chemical, and try to cut the risk of crop injury and costs. In 1967 and 1968, the USDA approved several chemicals for mixing, including Atrazine and Lorox, Atrazine and Ramrod, and Ramrod and Lorox. Manufacturers registered more chemicals for mixing in 1969, prompting Sylwester to comment that, "contrary to what some folks believe, these combinations are not being made to confuse you." Experts cautioned that farmers who did not purchase premixed herbicides needed to exercise caution when they mixed chemicals themselves in order to ensure that the herbicides were in the proper proportions.[36]

Mixing two or more herbicides was not the only way herbicide formulations became more complicated. Surfactants, or surface-active agents, helped the dispersal, spreading, wetting, sticking, and other surface-changing properties of the plants. Surfactants became popular in the 1960s and served as a way to get the most out of the herbicide. Some farmers used Atrazine with crop oil as a postemergence spray to increase absorption into the weeds. The oil not only helped the herbicide penetrate the leaf. Unlike a straight water mix, which evaporated quickly, the oil kept the unabsorbed chemical on the plant, gradually moving it to the base where it could be taken at ground level. Using surfactants gave farmers another means to obtain the most weed control from their increasing herbicide expense.[37]

The new herbicides were also riskier, however, since residues persisted longer in the soil and could remain into the next year and affect nontolerant crops. In 1949 weed scientists assured farmers that 2,4-D took from thirty to ninety days to dissipate out of the soil, which quieted any fears of the new technology. By the late 1960s, however, farmers had good reason to worry about residue. Sylwester noted that the 1968 crop suffered damage from 1967 Atrazine residues because of late season "rescue" applications compounded by dry conditions, which prevented chemical breakdown throughout the growing season and into the following spring. Studies indicated that Atrazine could remain active in the soil for much longer than 2,4-D, from four to twelve months depending on the amount applied per acre. Farmers could mix herbicides to reduce the risk of residues, but it required careful planning to ensure that the next year's crop could tolerate the selected herbicides. It was possible to mix two chemicals at a slightly lower application rate to lessen the chance that one of the herbicides would leave damaging residue for the next year's rotation.[38]

Some farmers and experts who experimented with herbicides hoped that chemicals could ultimately eliminate cultivation altogether. As early as 1949 an optimistic USDA weed scientist prophesied that farmers could eliminate cultivation with the aid of chemical weed killers. In the mid-1960s, some experts, farmers, and journalists began to consider minimizing their summer fieldwork, actually substituting chemicals for cultivation along with narrowing the rows in their fields. John Engelkes of Franklin County applied a heavy dose of incorporated Atrazine in 1965 on a germinating corn crop and avoided cultivating that season. He cultivated twelve acres as a control plot and noted that there was no visible difference in the quality of the crop between the herbicide-treated fields and the control plot. Although he did not weigh grain samples to get an accurate yield test, he was encouraged enough to try it again in 1966. Bob Gabeline of Louisa County found that with granule application of Randox he avoided cultivating on a portion of his acres, three-fourths of his corn needed only one cultivation, and he made only two passes with the cultivator through the rest of his fields. Herbicide costs were higher for this sort of replacement operation, but a Lucas County farmer noted that the cost difference was "more than paid for from yield increases and timeliness." He explained, "We have livestock and a considerable amount of hay. By not having to spend time cultivating beans, we gained extra time to take care of our corn, hay, and livestock."[39]

From 1945 to 1970 farmers who used growth-regulator herbicides realized only part of the optimistic promise of cheap and easy chemical control, but chemicals nevertheless became an indispensable element of Corn Belt agriculture. Despite some problems farmers embraced herbicides and actively worked to shape their use. The drive to reduce the labor and fuel expenses associated with cultivation provided farmers all the reason they needed to reorganize their work around herbicides. The advice from experts and manufacturers was important, but farmers themselves did what they believed necessary to remain in business, often ignoring the counsel of those experts.

Growth-regulator herbicide started as a practical and inexpensive solution to the postwar weed problem when farmers began using it on a trial basis in 1945, but ongoing efforts to incorporate this technology successfully into day-to-day practices changed the fundamental character of Iowa farming. The relationships between farmers, growth-regulator herbicides, and Iowa's farmland were constantly changing as farmers attempted to maximize production and deal with the unintended consequences of the new technology. As they grappled with the changing weed profile on their land, farmers found new ways to manage an old problem. A host of new chemicals and new techniques allowed them to maintain a degree of control over the grasses, but at a dramatically higher cost and degree of complexity. By the 1970s few Iowa farmers were debating whether to use herbicide; instead, they were debating which kinds to use and how to apply them. It was as important for farmers to know about herbicide as it was to know about tractors and corn. As Iowa farmers accepted herbicide and asserted initiative in using it, they combined it with chemical fertilizer and insecticide, increasing their commitment to corn and soybeans and livestock, reshaping the postwar farm landscape.[40]

Three

Fertilizer Gives the Land a Kick

In early 1944 a farmer from Poweshiek County noted that over the previous seven years commercial fertilizer had helped him get better yields on his pasture. "It gives the land a kick," he explained. Scientists and county extension directors agreed. They told farmers that in addition to pasture, hay, and small-grain acres, as much as half of the corn acres in the state could benefit from commercially manufactured nitrogen, phosphorous, or potassium to boost yields, depending on local soil types. Experiments from around the state showed that fertilizer stimulated root, leaf, stalk, and ear growth on corn plants and increased production of legumes such as alfalfa, too.[1]

Only a small number of Iowa farmers used commercial fertilizer on their crops of corn, small grains, hay, and pasture in the mid-1940s, however. Some Iowa farmers used fertilizers to compensate for deficiencies of their soil type. During the 1930s William Adams of Fayette County, for example, bought one or two tons of phosphate each spring. The sustained low commodity prices and subsequent low farm income of the 1930s made purchasing fertilizer a difficult or losing proposition for the majority of farmers, especially those who did not have any experience with commercial chemicals. Less than 5 percent of Iowa farmers in 1929 and only 3 percent in 1939 used fertilizer, a stark contrast to the practices of farmers in Corn Belt states to the east. In Ohio 62 percent of farmers and in Indiana 49 percent of farmers used fertilizer at midcentury. Most farmers in Iowa eschewed purchasing commercial fertilizer because they were successful in maintaining productivity with regular applications of manure and crop rotations of corn, small grain, and hay. The majority of farmers who used fertilizer specialized in crops with high labor requirements and high returns such as truck crops or vegetables. One observer, discussing the rapid spread of commercially purchased chemical fertilizer, claimed that in his boyhood people looked down on fertilizer as a "whip to the tired horse." The allusion to whips and tired horses implied that only farmers who failed to care for their land needed fertilizer, as only careless or cruel farmers lashed their horses. Successful crop-rotation strategies both balanced the draw on soil nutrients and limited populations of yield-robbing weeds and insects. Thrifty farmers also returned organic matter to the soil as livestock manure. For most Iowa farmers of the interwar period, fertilizer meant manure.[2]

After World War II, farmers in Iowa and across the country utilized commercial fertilizer as a regular part of their land- and crop-management routines for all their crops, especially corn and soybeans. From 1945 to 1970 the use of commercial nitrogen on American farms increased from 419 million tons to 7,459 million tons while the use of potash increased from 435 million tons to 4,035 million tons. Iowa farmers increased their total fertilizer use from 332,661 tons in 1950 to 2,333,411 in 1969 without significantly altering their total crop acreage. By the early 1970s farmers in pursuit of larger yields were using increasing amounts of fertilizer on their corn crop. They also successfully pioneered the use of nitrogen fertilizer for the soybean crop, helping increase the importance of soybeans in Iowa agriculture.[3]

Fertilizer Converts

Studies of fertilizer use indicate that the yield-boosting potential of commercial products was tremendous. Check plots on pastures showed that animals gained more weight grazing on fertilized pastures. One study showed that steers on pasture fertilized with phosphate applied at 155 pounds per acre gained 155 pounds per acre, while the steers on control plots only gained 105 pounds per acre. Cornfields in rotation treated with animal manure, lime, and 150 pounds of 0-20-0 fertilizer per acre yielded seventy-eight bushels to the acre, while fields treated with 200 pounds of 2-12-12 yielded eighty-five bushels of corn to the acre. These yields were remarkable considering that the average bushel-per-acre yield during the 1950s varied from the forties to a high of sixty-six in 1958. Throughout the 1950s and 1960s farmers and experiment station researchers regularly reported gains of ten or more bushels to the acre on fertilized fields.[4]

Young farmers were among the first converts to commercial fertilizer. In 1952, 56 percent of farmers under thirty-five years of age planned to use starter fertilizer (applied at planting time) on cornfields, followed by 50 percent of farmers aged thirty-five to fifty, and only 40 percent of farmers over fifty. Older farmers relied on the familiar system of crop rotation to balance nutrient use and used animal manure to put organic matter back into the soil. In 1957 the Jefferson County extension director reported that older farmers there would "probably never give in to fertilizer use, but the younger ones are showing increased interest." This generation gap continued into the 1960s, when another poll indicated that 71 percent of farmers aged thirty-five to fifty planned to use starter fertilizer on their 1963 corn crop and 66 percent of the group aged fifty through sixty-four would do so. Only 31 percent of farmers over age sixty-five planned to use starter fertilizer, with the youngest farmers trailing slightly behind the middle-aged leaders.[5]

Older farmers were not the only ones who were slower to take up chemical fertilizer. Farmers with smaller acreages lagged behind those with large areas under cultivation. As farmers employed herbicide to reduce labor demands at cultivating time and insecticide to limit yield losses due to insect infestation, and as they spread the cost of their machines over more acres, they wanted to ensure high yields by using fertilizer. Eight of ten farmers with more than 150 acres of corn planned to use starter fertilizer on their corn in 1963 while only 50 percent of those with less than 25 acres planned to use it. These farmers with small acreages might have owned more livestock, which enabled them to put more manure on their land per acre than those with extensive acreages. It is unclear if farms with small corn acreages were also farmed by older farmers who were less likely to use fertilizer.[6]

Regardless of age group or size of operation, farmers of the 1940s and 1950s had good reasons for using commercial fertilizer. The most compelling reason is that returns from fertilizer use generally exceeded costs. Of the 69 percent of Iowa farmers who used commercial fertilizer in 1953, 87 percent of them believed their crops benefited from the application. They saw for themselves that fertilizer worked on their own fields or on those of their neighbors. Over half the farmers surveyed reported it was the experiences of other farmers that had convinced them to try fertilizing. By contrast, only 20 percent of farmers began using fertilizer because of the efforts of the extension service or farm magazines. If there ever was a product that "sold itself" to farmers, it was commercial fertilizer.[7]

Farmers quickly accepted commercial fertilizer in part also because of their experiences with hybrid corn, a technological development they had adopted in the 1930s and 1940s. They learned they needed to fertilize the new hybrids to obtain maximum yields. Crops of high-yielding hybrid corn utilized a greater percentage of the available soil nutrients, drawing down supplies of nitrogen, phosphorous, and potassium at a faster rate than open-pollinated varieties. As farmers retired corn acres after 1933 as part of the Agricultural Adjustment Administration's corn-hog program, they offset declining production by using hybrid corn on their remaining corn acres. By 1942 all Iowa corn acres, regardless of farm size, were planted with hybrid seed corn. When the U.S. government lifted production controls during World War II, hybrid seed continued to be important, since government leaders wanted maximum production for the war effort and corn prices were high. The new hybrids were also suited to higher population densities, resulting in thicker stands of corn. The new strategies were to plant narrower corn rows and to grow more plants per acre. Both these developments made fertilizer an essential part of Corn Belt agriculture.[8]

Adapting Fertilizer to Conditions and Crops

Farmers accepted fertilizer as a valuable tool to boost yields and profits by raising more crops per acre, but how to use it was another matter entirely. Farmers confronted numerous issues, including which kind of fertilizer matched their local soil types, when to apply it, which tools to use, and even which crops responded best to chemical treatment. Experts labeled the process of balancing the draw down of soil nutrients and the application of fertilizer as the "mechanization of the nitrogen cycle." Fertilizer application could take place at various times throughout the year, and the popularity of each application technique varied. Some farmers submitted soil samples to the laboratory at Iowa State for testing, but even in 1962 farmers tested soil on only 20 percent of Iowa's cornfields. Farmers found that expert advice was frequently helpful, but not always. Many farmers experimented on their own to find out what worked best for their conditions or type of operation. Much of the discussion about fertilizers in the years up to the 1970s involved the timing and method of application.[9]

Farmers and experts agreed that applying starter fertilizer at corn-planting time allowed the plant to obtain a fast start. A farmer from Lyon County explained that he plowed under nitrogen and phosphate on his cornfields but also applied starter fertilizer. He noted that fast-growing corn outpaced weed growth, which allowed earlier cultivation or herbicide spraying. Garland Byrnes of Allamakee County recounted, "One year I ran out [of fertilizer] and left two rows without hill [starter] fertilizer. You could see a real difference in size of plants." In 1953 Ray Griffieon of Polk County claimed that using starter fertilizer increased his corn yield by ten bushels per acre, pushing his yield to a hundred bushels per acre, well above the average corn yields in the state.[10]

Throughout the 1950s and 1960s both farmers and extension professionals insisted that starter fertilizer made a substantial difference in yields. Even in the drought of the mid-1950s when many farmers cut back on fertilizer purchases, some farmers maintained that it was a good gamble. In 1956, at the height of the statewide drought, Oliver Hansen of Franklin County explained that starter fertilizer "pays well at least four years out of five. Sometimes, I get as much as 15 to 20 bushels of corn for only four dollars worth of fertilizer." According to this logic, not using commercial fertilizer was the real gamble and represented a failure.[11]

A key component of successful starter-fertilizer application was the location of the chemical in the seedbed in relation to the seed. According to a 1953 guide to fertilizer use, corn fertilizer needed to be at the same soil horizon (or level) as the seed, and about two inches away from it, in order to maximize the growing plant's utilization of the fertilizer. Later in the decade experts urged farmers to place the fertilizer lower in the seedbed and to the side. This prevented direct contact between the seed and the fertilizer at germination time and delayed germination. Instead the fertilizer was in position so that the first downward root shoots reached the fertilizer and, because it was not next to the seed, allowed for increased application rates. As one Webster County farmer stated, "We used to limit what we put on as starter. . . . But now with placement away from the seed we'll be using more." The use of starter fertilizer did not preclude using other techniques such as spring or fall plow down, as Lloyd Fosseen of Hardin County noted. "I think plow down followed with starter is the best," he said. "It gets the corn started and keeps it going." This type of combination was especially popular with farmers who applied fertilizer during the growing season.[12]

Starter fertilizer gained popularity in the 1940s through the 1960s. In 1951 only 38 percent of farmers reported using starter fertilizer, and 62 percent stated they did not use it. By contrast, in 1952, 48 percent of farmers planned to use starter fertilizer while 47 percent did not, a significant gain over the previous year. Just as striking was the conversion of the undecided farmers, who comprised 5 percent in 1951 but probably moved into the ranks of those who planned to use fertilizer, since there were no undecided farmers in the 1952 poll. By 1963, 65 percent of the farmers surveyed planned to use starter fertilizer. This trend toward starter fertilizer matched the recommendations of agronomists and farm journalists, who argued that tests on experiment station farms indicated that broadcasting fertilizer required almost twice as much fertilizer as using a planter attachment to get the same results in yields.[13]

Once the corn crop, boosted by starter fertilizer, was above the ground, many farmers applied nitrogen to the growing stand. A midseason application in June or July (called side-dressing) would place nitrogen next to the plants. In the 1950s, experts discussed the need for nitrogen during the growing season. One journalist explained, "Nitrogen is a growth-producing plant food something like protein for animals" and was necessary throughout the plant's life cycle. Farmers liked to see dark green cornfields, which indicated a healthy crop with adequate nitrogen. Farmers could side-dress nitrogen as they cultivated, with fertilizer applicators mounted on their implements to get the fertilizer beside the corn row, thereby making it available to the growing corn-root systems. Iowa State experiments in 1948 and 1949 showed that side-dressing 40 pounds of nitrogen per acre gave an average yield increase of thirteen and a half bushels per acre. From the perspective of farmers such as Bill Wilson of Keokuk County, "It really pays off. I get three dollars for each one [dollar] that I spend that way [side-dressing corn]." Side-dressing was like an insurance policy to help the growing crop reach maturity.[14]

Side-dress application of nitrogen was especially important on fields planted to second-year corn, the year before the field returned to small-grain and forage production in the rotation. Agronomists did not consider it necessary to fertilize first-year corn, which typically followed a legume-forage crop in the rotation, but they estimated that as much as 90 percent of Iowa's second-year corn would benefit from side-dressed nitrogen. Experts cautioned that this midseason application was not a replacement for proper rotation. Many farmers and most experts considered that planting corn on the same land for more than two years in a row was a poor practice, because it risked depleting soil nitrogen to such a degree that subsequent crops of nitrogen-fixing legumes in the rotation could not replace it. A growing number of Iowa farmers found that using nitrogen in combination with herbicide and insecticide allowed them to raise corn for three or more years, with some farmers even planting what became known as "continuous corn," in the same ground year after year. The continuous-corn strategy, promoted by Roswell Garst of Garst Seeds as a means of increasing farm profits, relied on heavy doses of fertilizer with nitrogen rates of up to 200 pounds per acre.[15]

Anhydrous ammonia quickly became the leading fertilizer for side-dress operations on Iowa cornfields. Anhydrous ammonia is a gaseous form of nitrogen that is stored under pressure and—during the 1950s—was injected into the soil at cultivating time. Unlike other sources of nitrogen, anhydrous ammonia dissipates when it comes in contact with air. At 82.5 percent nitrogen, it was the purest form of fertilizer available. In 1950 Elmer Carlson of Audubon County was one of the first farmers in the state to use anhydrous ammonia, but he was not alone for long. Farmers quickly adopted anhydrous ammonia for it required less handling than the bags of solid fertilizer. In 1961 Harold Whittlesey began farming in the northern part of the state, and he applied his first anhydrous ammonia in late May 1965. By 1968 Whittlesey had phased out his other fertilizer products and was applying only anhydrous ammonia. Whittlesey, like so many Iowa farmers, found it easier to hire a commercial applicator for this task. Hiring custom application permitted farmers to get on with other work during the busy month of June when they could make hay or take care of livestock. As a writer for *Wallaces Farmer* noted, "Frequently, custom applied nitrogen is cheaper than you can buy and put on yourself." Farm record books indicate that, by the 1960s, almost all farmers used anhydrous ammonia at some point in the year, often as side-dressing.[16]

Starter fertilizer and side-dressing with anhydrous ammonia gained popularity, but many farmers found that spring plow-down application met their needs. "Plowing down fertilizer" meant broadcasting it on the surface of the field and following up with plowing, which put the fertilizer deep in the seedbed. In the late 1940s, extension professionals hesitated to recommend plow-down application unless the farmer could use large amounts of nitrogen, since the principal advantage of plowing down fertilizer was getting plant food where it could be absorbed by the roots throughout the growing season. In late 1947 Ray Gribben, a farmer from Dallas County, noted in his diary that he was very satisfied with his corn yields on the land he had fertilized the previous spring.[17]

Spring plow down was especially valuable to those families with larger farms or with demands on their time from extensive livestock operations. Throughout the 1950s farmers extolled the virtues of plowing down fertilizer for corn because it reduced the workload at planting time. One Hamilton County farmer explained that his method of applying phosphorous and potash in February and nitrogen in March, just ahead of plowing on second-year-corn ground, was "a plain case of now or not at all." Without extra labor to help at planting time, to haul fertilizer and fill planter attachments, he was able to plant rapidly in May, a time when rain showers frequently disrupted fieldwork. The Iowa State agronomist H. R. Meldrum gave this practice a qualified endorsement, recognizing the labor shortage and noting that it was better than no fertilizer at all. Meldrum cautioned, however, that some of the fertilizer could wash away before it could be plowed under and that plow-down fertilizer was too deep to help get the young plants growing the way starter fertilizer did.[18]

Farmers continued to favor plow-down application in spite of expert advice to the contrary. In spring 1953 Bill Patterson of Madison County planned on plowing under fertilizer for the first time. He cited the hectic demands of June that kept him from getting into the field to side-dress corn with nitrogen. "Side dressing nitrogen on corn comes at the wrong time for me," he explained. "Corn, beans, hay and hogs all need attention at the same time. If I can broadcast nitrogen in April, it will save a lot of time during the busiest season." Plow-down application offered another perceived advantage, since the fertilizer was lower in the seedbed than it was with side-dressing. According to Walter Krauter of Lee County, plow-under fertilizer was "deep enough that corn roots get to it before the weeds. Some weeds never do get much of it [fertilizer] that way." Farmers such as Krauter turned the experts' advice of starter fertilizer upside down, arguing that shallow-placed starter fertilizer was available for the weeds, too. Proponents of plow-down fertilizer believed it was the best solution for the entire growing season and solved farm labor problems.[19]

In the late 1940s and early 1950s agronomists advised farmers to apply fertilizer in the fall as an alternative to spring plow-down application. They argued that just as much of the nutrients would be available the following spring as in fertilizer applied just before planting. In 1949 through 1951 extension agronomists initially promoted fall application of phosphate and potash for winter wheat and second-year pastures or hay ground. But in 1952 they began to promote fall fertilizing for corn. Extension agronomist Lloyd Dumenil explained that applying fertilizer before fall plowing made sense. He argued there was little chemical leaching or runoff of the fertilizer over the winter because precipitation was generally lighter in the fall than in the summer months. The biggest opportunity for runoff was in the spring, but most farmers applied fertilizer in the spring already, so there was little additional risk associated with fall application.[20]

Fall application also saved time in the spring when farmers were traditionally busy with seedbed preparation, hauling manure, and planting crops. This was especially important for farmers who expanded their acreage or dedicated a larger portion of their acreage to corn. In 1955 Weldon Franzeen of Greene County wrote about his experiences with fall fertilizing over the past two seasons. "Fall application does just as well as spring for me. I save a lot of work by having the fertilizer bulk spread." Franzeen explained that it was more economical to hire a commercial bulk spreader to do the job faster than he could do it himself. Gene Casey of Johnson County explained that fall application of nitrogen at plowing time saved a trip over the field in the spring, which reduced fuel costs as well as cutting the time needed to do fieldwork. "If you wait until spring," another Johnson County farmer cautioned, "you run the risk of weather, a shortage of time, and not getting the fertilizer [from distributors] when you need it." These farmers used fall application of nitrogen as a hedge against weather delays, the most volatile variable of farming.[21]

Farmers' testimony about the benefits of applying fertilizer in the fall rather than plowing down fertilizer in the spring included an appeal to the pocketbook that was hard to resist. Fertilizer distributors offered lower prices in the fall than they did in the spring, to ease problems with delivery and inventory management. A Humboldt County farmer found that a fall purchase saved "about $9 a ton when I shop around a little." Not all farmers agreed, of course. Carroll Swanson of Webster County stated he preferred to have the money in the bank rather than "in the ground." Yet the chance to buy at a discount and avoid incurring storage costs over the winter was attractive during times of cost-price squeeze.[22]

In the 1950s only a minority of farmers seized the potential cost and labor savings of fall fertilizing. In 1953, just one year after farm journalists and agronomists began to promote it actively, 17 percent of farmers surveyed in a Wallace-Homestead poll responded that they planned on spreading fertilizer in the fall for their 1954 crops. A dry autumn dissuaded some farmers from fall fertilizer, since they were less likely to plow because of the drought. By 1959, 18 percent of farmers applied fertilizer in the fall and spring, while 78 percent applied it only in the spring.[23]

In the 1960s, extension professionals urged farmers to use caution when applying nitrogen fertilizer in the fall. They offered new technical advice and caveats regarding fall application, including the risk of nitrogen loss. Agronomists expressed little concern about phosphorous and potassium loss, but this changed as fall application of nitrogen gained popularity. Agronomists issued their first admonitions about fall application of anhydrous ammonia in 1955. Since anhydrous ammonia was a gas that needed to be put into the soil lest it dissipate, it was critical to inject it into the soil at least four inches deep to prevent nitrogen loss. Injecting the gas at the proper depth allowed the nitrogen to bond with the soil.[24]

Controlling the depth of application was simple compared to other 1960s guidelines. In 1962 and again in 1964, extension agronomists backed away from a blanket endorsement of fall nitrogen applications. Agronomist John Pesek cautioned that "Fall applications of nitrogen are inferior to early summer applications, but we can't tell how much inferior they are." He advised farmers to wait until two or three weeks before the nitrogen was actually needed to avoid losses. Cooler soil temperatures prevented ammonium nitrogen from becoming a soluble nitrate, which was more likely to run off, a process called denitrification. Specialists urged farmers who applied fall nitrogen to wait until the soil temperature reached about fifty-five or fifty-seven degrees Fahrenheit for the 1962 season, revising those figures downward to between fifty and fifty-five degrees by 1964. In 1965 the recommended soil temperature was no more than fifty degrees. Extension professionals expected farmers to revise their techniques each year according to new research and guidelines.[25]

Denitrification became the subject of significant warnings to farmers in the late 1960s. One journalist informed his readers that the "risk of loss from fall application is smaller than once thought," because fall soil moisture was generally low. While he conceded that some nitrogen loss was possible, he emphasized that a gain of even one bushel per acre would offset a loss of up to one-fourth of the total nitrogen applied per acre. Another writer contended that fall application of nitrogen "got something of a black eye" in the 1967 growing season, since wet conditions that year led to increased denitrification. Extension agronomists hesitated to state exactly why farmers experienced unsatisfactory results with their fall-applied nitrogen, although moisture and temperature variations played a large part in the degree of failure. They insisted that fall nitrogen application at a depth of four inches could be successful as long as the soil temperature was below fifty degrees Fahrenheit. Regis Voss, extension agronomist, maintained that anhydrous ammonia applied in the fall was not likely to move.[26]

Neither experts nor farmers advocated abandonment of fall fertilizing, but publicity of the risks during the late 1960s raised awareness of the complexities involved in fertilizer use. The changing recommendations showed that expert advice, much like the skill and knowledge of the farmer, was subject to change as experiments and experience revealed new information. By the end of the decade, fall fertilizing had become an established practice, but one that required both precision and informed knowledge to ensure that as much of the fertilizer as possible stayed in the ground to begin the next growing season.[27]

Carryover

The concept that fertilizer remained in the ground from one year to the next was important throughout the postwar period. Regardless of the application method and timing, farmers and experts wanted to know how much chemical would remain in the ground to be available for the next crop year. Journalists first discussed the issue of carryover effects in 1952 when they cited an Iowa State study of nitrogen fertilizer from test plots across the state. Heavy applications of nitrogen plowed under on corn ground (180 pounds per acre of 33.5-0-0) yielded increases from between two to thirty-nine bushels to the acre over control plots, reflecting differences in soil type and moisture. The virtues of carryover became a constant refrain as experts and farmers discussed fertilizer use. Iowa State tests indicated that nitrogen carryover ranged from 5 to 50 percent of the previous year's application. In the early 1950s, experts calculated the carryover as the difference between the pounds of nitrogen applied per acre and the yield. For example, a farmer who applied 120 pounds of nitrogen per acre and had a corn yield of eighty bushels per acre could expect 40 pounds of unused nitrogen per acre left behind. More sophisticated analyses of the 1960s indicated that farmers who used less than 40 pounds of nitrogen per acre would not see much carryover, while those who used from 80 to 100 pounds of nitrogen per acre could expect to carryover from 25 to 35 percent of the nitrogen, with higher levels of application resulting in a greater percentage of carryover.[28]

Fertilizer carryover was especially significant during the mid-1950s as Iowa farmers struggled with drought. When 1954 began with a dry spring, Lloyd Dumenil, Iowa State fertilizer specialist, urged farmers to go ahead with their plans to use fertilizer, even if subsoil moisture was low. According to Dumenil, plants would run out of nitrogen in dry conditions. Supplemental nitrogen would allow the crop to survive longer, perhaps until the next rain. In extreme drought, plants utilized little of the available nitrogen fertilizer. As a writer for *Wallaces Farmer* argued, "If plants don't use the fertilizer this year, most of it will stay in the soil for the next crop." Fertilizer promoters emphasized that it was still economically worthwhile to fertilize, even though the payoff might come in the next year.[29]

In spite of the encouragement from fertilizer experts that applying fertilizer in dry years was profitable, Iowa farmers had their own ideas. After using a record 563,000 tons of commercial fertilizer in the state in 1954, farmers cut their usage in the dry years of 1955 and 1956. In 1955 farmers reduced their fertilizer use by 55,000 tons. With fears of continued dry weather in 1956 and 1957 farmers adjusted their fertilizer consumption accordingly. They used less than 400,000 tons both years, a 20 percent decrease from the 1954 high. One Clay County farmer used starter fertilizer and 75 pounds of anhydrous ammonia per acre on a dry portion of his farm in 1957, but his corn yielded only sixty bushels to the acre that year when it normally yielded eighty bushels. "I couldn't see much return from fertilizer," he concluded. A county extension director observed that the fertilizer cause "suffered a serious setback" during the drought of 1954–1956. Carl Peterson of Palo Alto County boosted his 1955 fertilizer use over the previous 1953 and 1954 levels, possibly following the lead of experts who asserted that fertilizer in dry years could pay. However, like many Iowa farmers, he scaled back in 1956. In 1957 he did not use any fertilizer for the first time since 1944. His actions reflect the retreat from fertilizers from 1955 to 1957. Only the return of normal rainfall in 1958 prompted farmers to resume aggressive fertilizer use. That year farmers used almost as much commercial product as they did in 1954.[30]

Fertilizer experts failed to convince Iowa farmers that fertilizer and carryover paid dividends in drought years, but this did not stop them from defending their position. In 1958 Al Bull of *Wallaces Farmer* wrote about farmers who had good results from using fertilizer in the midst of drought. The story of Gerald Pedersen from Clarke County indicated the tension over fertilizer use as well as the potential for profit. In 1957 Pedersen had rented some land in addition to what he owned. The landlord did not want to fertilize but agreed that Pedersen could fertilize his half of the rented ground, presumably on the assumption that the tenant would bear all the risk if it was a bad year and still reap the benefit from any yield increase. Pedersen applied starter fertilizer and side-dressed with nitrogen. "All summer," Pedersen noted, "you could see to the row where fertilizer had been applied. The landlord's corn without fertilizer made forty-five to fifty bushels per acre. My corn went seventy to seventy-five bushels." In Bull's telling of the story, the cautious or parsimonious landlord lost and the risk-tolerant and innovative renter gained one-third higher yields, which suggested that the farmer who stayed the chemical course would prevail. Most farmers disagreed and chose to save their money for a year when the fertilizer might provide maximum return on the investment.[31]

Fertilizer for Soybeans—Farmers in the Vanguard

Drought was not the only circumstance that caused farmers to limit their fertilizer use. Until the 1960s most experts and farmers considered that applying nitrogen fertilizer to soybean fields was wasteful, in part because they believed soybeans might benefit from any carryover fertilizer. Experts believed that soybeans did not respond to fertilizer as well as corn did, unless the land was especially low in potassium or phosphorous. Since fertilizer was one of the biggest crop expenses, extension advisors counseled farmers to fertilize the "high-value crop," which was corn on most Iowa farms. D. L. Armann of Polk County reflected this view when he planted his soybean crop on land that had been in forage. He claimed he could "profitably fertilize corn but not beans so I let beans harvest sod nutrients." Extension professionals and farm journalists discouraged farmers from fertilizing the bean crop since it was unlikely that the benefits of fertilizer for soybean fields outweighed the costs.[32]

Farmers who fertilized their soybean crops challenged the experts' wisdom. Alfred Accola of Polk County fertilized his soybean crop in 1962 and obtained over sixty-three bushels per acre in a test plot. In 1966 growers who fertilized soybeans and obtained top soybean yields gained attention in the farm press. A survey of Iowa farmers who raised high-yielding soybean crops, conducted in 1967 by the National Soybean Improvement Council, showed that fertilizer was an important part of successful soybean growers' techniques. Roger Harms, a farmer from Butler County, entered a 1967 yield contest sponsored by American Cyanamid Company. He applied 600 pounds of fertilizer per acre and produced eighty-five bushels to the acre on a five-acre check plot and seventy bushels per acre on a fifteen-acre field. Harms noted that he had been fertilizing beans since 1962 and gradually increased the amount each year. While the fertilizer bill for 600 pounds per acre was high, the increased yields allowed him to net $150 per acre on the five-acre plot, which more than offset his expenses for seed, fertilizer, herbicide, machinery, and labor.[33]

The experts belatedly agreed that fertilizer could make a difference on the soybean crop, although none of them endorsed using the high levels of fertilizer that Harms did. One Iowa State agronomist contended that soybean yields of fifty bushel to the acre "are common enough to indicate this may be a good goal to shoot for" and that fertilizer could help farmers reach that goal. The same type of crop management that farmers used for the corn crop could work for soybeans. Iowa State agronomists cautioned that soybeans would not show the same type of dramatic increases as corn but conceded that not fertilizing beans "has been holding back yields." In 1968 Iowa State soybean fertility specialist C. J. De Mooy noted that, contrary to earlier beliefs, soybeans were a deep-rooted plant, which meant that deep application of fertilizer could be useful to the plant later in the growing season when the root system was more fully developed. Other experts echoed this, arguing that fertilizer could increase yields enough to be profitable.[34]

Runoff

Critics outside of agriculture pointed out that there were costs associated with fertilizers other than just financial ones. Just as public concern had mounted over the consequences of pesticide use, the idea that fertilizers could be harmful gained momentum in the 1960s. Rachel Carson's *Silent Spring* was an attack on the indiscriminate use of pesticides, but the implications for discussions of water quality were clear. In Carson's view, farmers, led by the USDA and chemical manufacturers, used chemicals in such a way as to threaten the survival of wildlife and people. Carson attacked those who would control or subjugate nature with chemicals, and she singled out scientists and bureaucrats at the USDA's Agricultural Research Service for special criticism. As historians have noted, Carson's indictment found a receptive audience during an era when people perceived many parts of the environment as becoming increasingly toxic, with concerns ranging from nuclear fallout and suburban septic tanks to pesticides in forests and suburban lawns. In 1965 Congress passed the Water Quality Act, which authorized the secretary of health, education, and welfare to formulate water-quality standards for states in the absence of state action. More than any previous federal legislation, the 1965 act put water quality on the national agenda.[35]

In Iowa in the 1960s there was indeed evidence that fertilizer runoff was becoming a public health problem. Water samples submitted to the state hygienic laboratory from private and rural water supplies indicated high levels of nitrate concentrations. High levels of nitrates in the water were new in the 1960s, but there was a difference in the type of wells sampled. By the 1960s there were fewer shallow hand-dug wells in use as many Iowans had invested in new deep wells. These deep wells were contaminated with nitrates to the same extent as the old-style wells, suggesting that nitrates percolated deeply through soil and bedrock and that the contamination problem was severe. Experts were cautious not to attribute high nitrate levels to the increased use of nitrogen fertilizer since 1945, but fertilizer was probably the principal reason for current levels of nitrates in well water.[36]

Few Iowa farmers expressed concern about the risks of fertilizer runoff and the environmental costs of fertilizer use, but there was a sense among farmers, journalists, and agricultural scientists that public criticism and pressure could potentially force Congress or state legislatures to act in ways that would impede farmers' ability to make a living. Seeley Lodwick of Lee County and president of the American Soybean Association was one of the few farmers who publicly expressed concern about the effects of pollution on farm practices. In 1968 Lodwick editorialized that "Even tho these pollution control laws are not fully enforced today, we have every reason to expect mounting pressure for clean, fresh water from the eighty percent of our population who live in the cities of Iowa. These laws," he argued, "will someday—soon—be enforced." In 1970 Al Bull asked, "Are fertilizers polluting our streams?" He recognized agricultural scientists' understanding that too much nitrogen runoff could be a problem and that there was evidence of nitrate contamination of water supplies. Bull quoted Regis Voss of Iowa State, who conceded (albeit torturously), "I'd certainly not say that fertilizer isn't contributing to nitrate content of surface waters." Voss and other scientists refused to blame farmers who used nitrogen fertilizer according to recommendations, even as they acknowledged that farmers who applied nitrogen improperly or excessively contributed to pollution problems. Experts argued that fertilizer actually helped reduce pollution by preserving organic matter in soils and preventing erosion. Bull and the experts understood that nitrogen fertilizers could be pollutants, although, as Bull stated, the real problem was "unrealistic ecologists" who had "visions of crystal clear streams," which, he argued, had never been a part of prairie ecosystems. Insiders and experts feared that ignorant outsiders might mobilize public opinion and restrict farm fertilizer use based on unrealistic assumptions. The best result farmers could hope for was to keep idealistic scientists and the public from pressuring lawmakers into enacting "unreasonable restrictions" on nitrogen use.[37]

In the early 1970s, however, ecologists and concerned citizens had the clout to change the nation's laws regarding water quality. Many people saw the environment as increasingly toxic and blamed technology. Activists such as Barry Commoner testified about the nitrate levels in neighboring Illinois streams, while state pollution-control boards across the nation focused attention on nitrates, soil erosion, and threats to wildlife through chemical poisoning and loss of habitat. In 1972 Congress followed the 1965 Water Quality Act with the Federal Water Pollution Control Act, known as the Clean Water Act, to preserve the "chemical, physical, and biological integrity of the nation's waters." The authors of the Clean Water Act focused on pollution from point sources such as factories, not farms. Still, for the first time, outsiders and non–farm related interest groups were scrutinizing rules for chemical fertilizers. Iowa state's Chemical Technology Review Board (created in 1970) considered it important to study the movement of agricultural chemicals into ground- or surface waters but did not recommend any legislative action in 1970. Farmers would not be required to change their practices in 1970, but the Iowa legislature and the U.S. Congress both recognized the legitimacy of the concept of ecosystems, which would change the approach of scientists, policy makers, and farmers in the coming years.[38]

In the years up to 1972 most Iowa farmers in their discussions of fertilizer addressed potential yield gains, expenses, and methods or timing of application, not the ecological costs or environmental damage. For most farmers chemical fertilizers—especially nitrogen in the form of anhydrous ammonia—helped them make money. Fertilizer runoff and the pollution of ground- or surface water were not major concerns. While there were farmers who recognized that fertilizers could be harmful to themselves and others, they observed little evidence that fertilizer technology was bad for them or their land. As far as they knew, experts and innovative farmers told the truth when they claimed fertilizers increased yields and offset any financial costs: fertilizer gave the land a kick.

From "Whip to the Tired Horse" to Good Management

It is difficult to know the degree to which farmers profited from using fertilizer on their corn, soybean, and forage crops. It is clear that they believed fertilizer helped. Within just a few years after the war they had made it a regular part of their program. A poll of Iowa farmers in 1949 indicated that approximately 43 percent of farmers used fertilizer, and most of them applied it to cornfields. In 1959 approximately 68 percent of farmers used fertilizer, even though less than half of the farmers surveyed conducted soil tests to determine what kind of fertilizer they needed and how much to apply. A 1969 survey showed that 96 percent of Iowa farmers used fertilizer. Those who applied commercial fertilizer were pleased with the results, with the exception of the 9–14 percent of farmers who reported no effect from their fertilizer during the drought years of 1954 and 1955. This short period of drought was only a temporary reversal for fertilizer use.[39]

Farm records from this period provide a closer look at the general pattern of commercial fertilizer consumption. Farmers who began farming in the 1930s were less aggressive about their fertilizer use. Carl and Bertha Peterson of Palo Alto County first purchased phosphate in 1945 and from 1945 to 1951 continued to buy one or two tons per year at an average cost of sixty-four dollars per year. Then the Petersons made a major commitment to fertilizer in the years from 1952 to 1956. They purchased high-analysis fertilizer with a high phosphorous content almost every spring, and starter fertilizer for corn containing the insecticide Aldrin in the summer, and they occasionally made fall applications. They used approximately the same type and amount of commercial product every year, applying 5-20-20 and 5-20-10 starter fertilizer every year from 1954 to 1966. In subsequent years their records become less precise in terms of what kind of product they used. At the end of the 1960s they purchased approximately the same amount of product as they previously did with some variations in timing and composition.[40]

The Petersons were the exception to the rule of increasing fertilizer use. Most Iowa farmers increased their use of commercial fertilizer in hopes of obtaining bigger yields. In the 1960s Iowa farmers doubled the amount of nitrogen they used per acre, from 45 pounds per acre in 1964 to 104.3 pounds per acre in 1969. This type of growth was in line with changes in other midwestern states. Minnesota farmers nearly tripled their application rate, and Illinois farmers increased their rate by over one-third. Rudolf Schipull of Wright County was in step with the trend as he increased his fertilizer use in the late 1960s. From 1963 to 1965 he applied an average of 17 tons of fertilizer each year on the two farms he operated. From 1966 to 1970, however, he applied an annual average of 29 tons on those two farms. Farmers who harvested the largest corn crops used the most fertilizer. A farmer from northeast Iowa who harvested a 125-bushel-per-acre corn crop stated that his neighbors could have obtained similar yields "if they'd use enough fertilizer." Extension professionals urged farmers who wanted to grow corn that yielded from 100 to 120 bushels to the acre to use from 120 to 150 pounds of nitrogen and a maximum of 37 pounds of phosphorous and 65 pounds of potassium on each acre. As farmers used more fertilizer per acre over the course of the 1950s and 1960s, fertilizer became the leading crop expense, even as costs per pound declined. Fertilizer comprised 24 percent of total costs of raising the corn crop in 1958, calculated on a per-acre basis, and 39 percent of total costs in 1967.[41]

Perspectives on what constituted good farming changed as older farmers and many of those with small and more moderate-sized farms quit farming. Fertilizer, previously viewed as the whip to the tired horse used only by inferior farmers, was now perceived as a legitimate, necessary technology that characterized good management. The change in commercial fertilizer use from 182,651 tons in 1946 to 2,648,196 tons in 1970 was a demonstration of the confidence the majority of farmers had in fertilizer technology. In 1971 Iowa farmers used fertilizer on 95 percent of all corn acres in the state. They timed applications on crops, especially cornfields, according to their other labor demands, cut their fertilizer use during drought conditions in spite of expert advice to the contrary, and rejected expert advice that using fertilizer on soybean fields did not pay. As the 1970s began, however, it had become clear that fertilizer also gave the land and water a kick in a negative sense. In the years to come, farmers and agricultural experts were aware that fertilizer use required just as much caution as using pesticides.[42]

Broken stalks in this Butler County cornfield show the extent of damage from European corn borer infestation in 1948. Photo taken by Harold Sherma, Bristow, Iowa. *Annual Report, Entomology,* 1948. Iowa State University Library/ Special Collections Department. Reproduced with permission.

68

below—The photo caption from an extension entomology report reads "Dusting for Corn Borer Control." In the late 1940s and early 1950s farmers used DDT for midseason rescue operations like this. By the late 1950s, however, they relied less on DDT and more on selecting varieties of borer-resistant hybrid corn. *Annual Report, Entomology,* 1948. Iowa State University Library/Special Collections Department. Reproduced with permission.

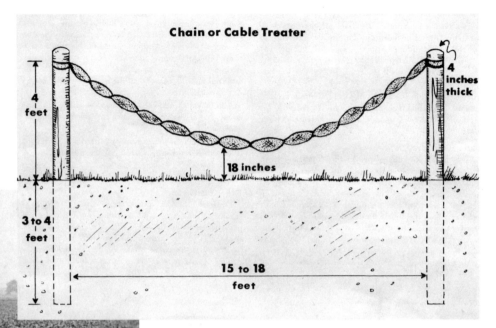

Chain or Cable Treater

4 feet

4 inches thick

18 inches

3 to 4 feet

15 to 18 feet

above—Farmers installed devices for horn fly control in their pastures or feedlots. The chain or cable was wrapped in burlap soaked in DDT for beef cattle or methoxychlor for dairy cattle. "Flies on Livestock," Agricultural Extension Service, Iowa State College, Ames, Iowa, *Pamphlet 200,* May 1953. Iowa State University Library/Special Collections Department. Reproduced with permission.

Extension entomologist Harold Gunderson and Earl Burbridge of Mondamin searched for soil insects in a field treated with insecticide, May 19, 1955. *Annual Report, Entomology,* 1955. Iowa State University Library/Special Collections Department. Reproduced with permission.

Spraying herbicide such as 2,4-D along fence lines at the edges of fields was an important part of the war on weeds. Weeds that spread into cropland and pastures through rhizomes or dispersing seeds cut crop yields. This photo of Orville Wilson from Dallas County was taken in July 1950, although the editors of *Wallaces Farmer* used the image in the June 16, 1951 issue. Reproduced with permission.

above—This photo of Bernard Derner of Dickinson County was originally titled "Spray *vs.* Weeds." Derner used dropped nozzles suspended from the spray boom to get the chemical onto weeds below the corn leaves. Farmers often substituted spraying for cultivating. *Wallaces Farmer*, July 10, 1963. Reproduced with permission.

below—Almost all Iowa farmers were using herbicide on at least a portion of their farms by the mid-1960s. As cartoonist Hank Warner suggested, knowledge of herbicide was essential to becoming a farmer in the postwar Corn Belt. *Wallaces Farmer,* June 5, 1965. Reproduced with permission.

"Dear, our son is going to be a farmer. He said his first words today: "Atrazine—Treflan—Dowpon—2, 4-D—Ramrod—Knoxweed and Alanap!"

below—Side-dress fertilizer applied on a Cherokee County cornfield on June 28, 1949. The owner of this farm reported that he did not use fertilizer on his fertile bottomland fields, only on his upland cornfields. *Wallaces Farmer,* June 17, 1950. Reproduced with permission.

above—Farmers boosted yields by using anhydrous ammonia, a gaseous fertilizer that was 82.5 percent nitrogen. Charlie Knudsen of Audubon County is pictured here, injecting nitrogen in June 1951. Knudsen reported he would use his applicator on 65 of his own corn acres and 1,600 acres for other farmers on a custom basis in 1951. *Wallace Farmer,* May 2, 1953. Reproduced with permission.

above—A commercial applicator spreads 0-20-20 fertilizer on oat stubble on the Russell Gould farm in Worth County in this undated photograph. *Wallaces Farmer*. Reproduced with permission.

below—In the early 1970s farmers faced increasing restrictions on the use of feed additives such as growth hormones and antibiotics. This cartoon by an Iowa State University student suggests that the rules were becoming more complicated. Some Iowa farmers followed the new regulations, but others could not or would not do so. *Iowa Agriculturist,* Spring 1972. Iowa State University Library/Special Collections Department. Reproduced with permission.

Modern milking parlors like this one allowed farmers to work more efficiently and with more comfort than in older stanchion barns. Jim Depenbusch (pictured here) and his father milked a fifty-cow herd when this photo was taken in April 1964. *Wallaces Farmer*. Reproduced with permission.

Automated feeding with fence-line feed bunks saved backbreaking labor, opening and closing gates, and getting stuck in muddy lots. Fred Hamilton of Pottawattamie County used this automatic unloading feed wagon in 1967. *Wallaces Farmer,* February 11, 1967. Reproduced with permission.

Making hay was a two-man job with the self-tying hay baler that was commercially available in 1944. *Wallaces Farmer,* June 24, 1958. Reproduced with permission.

Farmers who specialized in livestock production could make high-quality forage by feeding chopped hay directly from the wagon or storing it as haylage. Gilbert Hoch, a Marion County hired man, is pictured chopping hay in 1950 on a farm near Knoxville. *Wallaces Farmer and Iowa Homestead,* June 2, 1951. Reproduced with permission.

above—Iowa State College engineers designed self-feeding hay keepers to cut labor costs by reducing the amount of feed farmers handled for beef and dairy herds. This 1956 photograph of a dairy farmer, cows, and hay keeper is from Carroll County. *Wallaces Farmer.* Reproduced with permission.

below—Newal Foust of Floyd County cut his oats with a windrower in preparation for combining in 1952. Farmers who combined their oats sometimes used a windrower a week to ten days before combining in order to let the grain cure in the field. *Wallaces Farmer,* August 1952. Reproduced with permission.

above—A pickup attachment on a tractor-drawn combine lifts the windrowed oats onto the combine to be threshed. Farmers who used combines for their small-grain harvest abandoned the labor-intensive work of community threshing rings. *Wallaces Farmer,* June 20, 1959. Reproduced with permission.

below—Combines gained popularity since they could be used for small grain as well as soybeans. This Hardin County farmer combined soybeans on his own farm and as a custom operator in 1962. Harvesting at night during good weather was preferable to waiting for daylight and risking the loss of bean pods in a severe storm. *Wallaces Farmer,* October 5, 1962. Reproduced with permission.

above—Farm families built many new corncribs in the 1940s and 1950s to deal with increased yields from hybrid seed, fertilizer, and pesticides. Cribs like this one in Washington County held ear corn harvested by mechanical pickers. Changes in harvesting soon made these cribs obsolete. A. M. Wettach Collection, State Historical Society of Iowa, Iowa City. Reproduced with permission.

below—Farmers across the state constructed inexpensive temporary corncribs from snow fence to store ear corn harvested by mechanical pickers. This roofed structure in Cerro Gordo County was more elaborate than most snow-fence cribs. *Wallaces Farmer,* August 20, 1955. Reproduced with permission.

In 1953 Homer Bugby of Dallas County used his picker-sheller to harvest ten acres of corn per day on three farms: his son's, his son-in-law's, and his own. Picker-shellers allowed farmers to harvest earlier and to avoid problems such as losing ears to ear drop from European corn borer damage or inadvertently shelling part of the crop in the picker. *Wallaces Farmer,* November 21, 1953. Reproduced with permission.

left—Portable crop dryers like this Behlen model (cutaway view) made it possible for farmers to use picker-shellers or combines for corn, since they could dry the grain to prevent spoilage in storage. Behlen Manufacturing Company. Reproduced with permission.

bottom—Corn did not always dry evenly in the new grain bins that displaced corncribs. In 1962 Eugene Sukup, a farmer from Sheffield, designed a stirring device like this 1964 Stirway model to break up pockets of wet grain, which could ruin the contents of a bin. Sukup Manufacturing Company. Reproduced with permission.

John Deere was the first company to introduce a combine with an attachment for harvesting corn. Farmers who purchased this kind of machine converted old cribs to granaries, constructed cylindrical grain bins, or even sold corn directly from the field. Reproduced with permission.

Four

Feeding Chemicals

Veterinary care, commonly known as "stock doctoring," has long been concerned with healing sick or injured animals. Good farmers do their best to prevent injury to livestock or outbreaks of disease, but there have been greater limits to what farmers and veterinarians could do. During the early twentieth century, however, new serums and inoculants were introduced to reduce mortality and improve herd health by battling diseases such as hog cholera, tuberculosis, and erysipelas. These drugs helped farmers deal with acute conditions, that is, diseases already present in or that threatened the herd.

After World War II, drugs proved useful in accelerating the livestock production cycle. Two new kinds of powerful drugs came into use: antibiotics and growth hormones. Farmers could use antibiotics to control outbreaks of disease and for another, more surprising application. Animals that consumed feed mixed with antibiotics at subtherapeutic levels reached market weight with less feed. Similarly, synthetic growth hormones stimulated rapid growth and conversion of feed to living tissue. Since feed costs represented a large portion of the total cost of raising an animal from birth to market, innovations that maximized production while cutting costs were very attractive to farmers confronting the cost-price squeeze and rising labor costs.

Both these new pharmaceuticals were developed in the 1940s and 1950s. Researchers used a synthetic form of estrogen, the growth hormone diethylstilbestrol (hereafter DES or stilbestrol), in dairy research during World War II to return dry milk cows to production in order to meet high wartime demands for dairy products. In 1946 farmers learned that penicillin could restore production to cows infected with mastitis, an infection of the udder that slowed or stopped milk production. Tests conducted at Purdue University in 1947 and 1948 indicated that a pellet of stilbestrol could be injected under the skin of a heifer to promote rapid weight gain.[1]

In 1951 Wise Burroughs, a scientist at Iowa State, observed that lambs fed with hormone-laced feed gained weight faster than those on a normal ration. Although researchers conducted the first feeding experiments on sheep, cattle were a logical choice for experimentation, since the period from birth to marketing was measured, not in months like hogs and sheep, but in years. In 1954, as a result of research performed by Burroughs and others at Iowa State, feed manufacturers incorporated stilbestrol into cattle feed. Cattle were a natural choice for the commercial introduction of this substance because consumers favored beef and had lived with less of it during the hard times of the 1930s and the rationing of World War II. Beef consumption surged in the postwar period while consumption of mutton and lamb remained stable.

The origin of the growth-enhancing properties of antibiotics was more obscure than that of hormones. In the 1940s scientists recognized that B_{12} had growth-stimulating properties when fed to livestock. They called the new vitamin mix Animal Protein Factor (APF). Farmers who used APF found not only that the substance also prevented swine dysentery (commonly known as bloody scours), a leading killer of young pigs, but also that their hogs gained weight 10–20 percent faster than animals without it. In 1949 E. L. R. Stokstad and T. H. Jukes discovered that only B_{12} that had been produced as a by-product of antibiotic manufacturing possessed this quality. It was the residue, they concluded, not the vitamin itself that was the APF ingredient that actually stimulated growth. In 1950 feed manufacturers began adding several kinds of antibiotics to feed, including Aureomycin and streptomycin.[2]

"Wonder Drugs," Extension, and Farmers

Journalists and scientists labeled antibiotics "wonder drugs" that could solve the problem of high infant mortality among hogs and accelerate the growth rate of feeder pigs destined for market. Tests conducted at experiment stations and laboratories around the country indicated that hogs and other animals benefited from rations that included antibiotics. A writer for *Wallaces Farmer* reported that Hormel Institute researchers in

Minnesota discovered runty pigs fed with Aureomycin mixed into the feed could catch up with healthy pigs. Furthermore, healthy pigs that received feed with antibiotics grew twice as fast as pigs on a normal ration. Damon Catron and his Iowa State colleague claimed that this was "New Hope for 20 Million Runts" across the United States, since one pig out of every litter was abnormally small or unhealthy. They explained that farmers who used antibiotics could expect "faster gains, more pork from 100 pounds of feed, and less trouble from scours." By feeding 5 milligrams of Aureomycin per pound of feed farmers could produce 100 pounds of pork with 349 pounds of feed instead of 369 pounds.[3]

Advertisers promptly featured the growth-enhancing properties of antibiotics with testimonials from farmers. In 1952 the Pfizer Company touted the remarkable gains obtained by a farmer from eastern Iowa who added Terramycin to his hog rations. He marketed 206-pound hogs when they were 139 days old. These animals consumed 3.4 pounds of feed for every pound of weight gain, costing 14.8 cents per pound of pork. The following year another Pfizer advertisement provided context for those numbers. Pfizer claimed that 50-pound hogs fed with Terramycin in their ration reached market weight in 94 days, compared to 111 days for hogs without the antibiotic-enriched ration. Using the antibiotics enabled farmers to reduce the number of days on feed and save at least 10 percent of feed costs.[4]

Farmers read the news and wanted the drugs. In response to publicity in the farm press in the late 1940s, hundreds of farmers sent postcards and letters to Catron and Iowa State to obtain both information about APF and even the actual medication. The following exchange was typical of the correspondence between farmers and Catron in 1950 and 1951. "Dear Sirs: I am a young hog raiser up hear at Iowa Falls and am interested in learning all I can about getting some off the new wonder drug 'aureomycin.' Can you inform me as to what I can best do to get some? Also how should I feed it." Catron responded that it was premature to bestow wonder-drug status on Aureomycin. He also noted that pure antibiotics were available only by prescription. Only certain APF concentrates contained the antibiotic residues, and as of mid-1950 there was no way to get the drugs.[5]

Inquiries continued into 1951. The Wright County extension director wrote to Catron in April and reported that area dairy farmers who fed skim milk to their hogs as a supplement wondered how to get antibiotics into the milk. Catron assured him it was possible to combine the commercially available antibiotic-vitamin premix with skim milk, since the antibiotics and the vitamins were all water soluble. Wright County farmers who attended a swine producers' meeting also wanted the updated formula for swine rations in order to take advantage of the latest feeding developments.[6]

Farmers who used antibiotics did not always get the expected results. In 1951 a farmer from Woodbury County complained that he had used commercial hog feed containing vitamin B_{12} and antibiotics, but his herd still presented symptoms of bacterial infection and nutritional deficiencies such as scouring, or diarrhea. A Polk County farmer who had a history of swine dysentery in his herd reflected, "I thought the antibiotics I've been feeding would control this." The problem for both these farmers was that the antibiotic-enriched rations they were using could not overcome an inherent and common problem in farm management of the mid-twentieth century. Many farmers used the same hog lots year after year, which provided an optimal environment for the growth of parasites and bacteria that sickened animals, caused death, or at the very least slowed their growth. Extension directors and Iowa State faculty urged farmers not to use antibiotics as a substitute for good management. As the experiences of the two hog raisers from Woodbury and Polk counties indicated, farmers needed to do more than simply provide feed with antibiotics. Rotating animals through different lots from year to year was an ideal solution, since it prevented the buildup of dangerous pathogens in the soil. Other solutions included the cleaning and disinfecting of hog houses and shelters. For those who wanted to use drugs, it was important to adjust the dosage to meet the specific needs of each farmer's herd. Catron noted that different quantities of B_{12} and antibiotics were needed for healthy animals than for sickly ones.[7]

Some farmers experimented with injecting antibiotic pellets into hogs. If a pellet of Bacitracin was implanted under the skin behind the ear when the pigs were just two days old, the animals would get the growth-promoting benefit of the antibiotic before weaning. When Victor Nicolet of Cerro Gordo County moved to a new farm in the spring of 1952, he weaned only sixty-four of the seventy-two pigs he farrowed. This loss of over 10 percent of his pigs was only the beginning of his bad news. The surviving pigs gained weight slowly throughout the year. Nicolet learned the hard way that the hog lots on his new farm were infested with disease. After injecting his pigs with antibiotic pellets that fall, Nicolet used the same rations and lots and fared much better in terms of rate of gain and herd health. While he did not leave any pigs untreated as a control, he believed that the treatment made a difference. His brother and father were convinced, too. They injected their pigs in 1953, basing their decision on the success of Victor's experiment. "The cost is low," Nicolet claimed. "So the treatment doesn't have to do much good to be worth more than it costs." Injecting antibiotic pellets, however, was more work than feeding commercially prepared feeds with antibiotics included in the ingredients.[8]

Antibiotics played an ever more important part in the changing nature of hog production in the 1950s. Experts and farmers attempted to get hogs to market weight faster in order to cut costs. The sooner farmers could wean young pigs from sow's milk to solid feed, the sooner those pigs could get to market and the sooner the sows could become pregnant again. Antibiotics were invaluable in speeding up the process. Farmers who raised hogs had traditionally weaned pigs at approximately eight weeks of age, relying on the sows to provide the calories, nutrients, and antibodies while the young animals gradually learned to consume grain or forage. At weaning time, some farmers provided starter feed that often included molasses or other sweeteners to help the young pigs make the transition. In the 1950s Catron and other experts promoted the replacement of sow's milk with carefully balanced feeds that enabled young pigs to grow faster than they would on sow's milk. This also prevented sows from losing weight through the lactation period and permitted them to reproduce sooner. Early experiments focused on the use of synthetic milk that included antibiotics, but there were problems with getting the milk to the pigs. The liquid required careful mixing, it was easily wasted through spillage, and easily frozen in cold weather.

Beginning in 1954 experts at Iowa State introduced a new feed ration called "pre-starter" to aid with early weaning. It was a transition feed to move week-old pigs from sow's milk to starter feed. Labeled "I.S.C. Pre-starter 75," the formula included antibiotics just like the synthetic milk, but this new feed was in dry form, which was easier to handle and feed. The promoters of pre-starter argued that traditional weaning cost $6.29 to feed one pig and the lactating sow for eight weeks. With pre-starter it cost only $5.47 to feed the pig for eight weeks and its mother for two weeks. The cost savings of $0.82 multiplied by dozens or even hundreds of pigs per year amounted to hundreds of dollars that the farmer could realize by early weaning. Catron and other swine nutritionists cautioned that this practice was not for every hog producer. Farmers who were committed to a high degree of management in sanitation, housing, and feeding were best suited to early weaning.[9]

Farmers flooded Damon Catron with questions about "Pre-starter 75," subsequent revised formulas, and new mixes that included antibiotics such as "Plus" and "3-Nitro." Again, hundreds of inquiries arrived from across the state and around the nation, from individual farmers, county extension directors, and farm managers for Farmers National Company and Doane Agricultural Service. Catron provided copies of the ration formula for farmers to mix themselves or to provide to local feed millers. Catron encouraged farmers to purchase feed from reputable dealers such as the Carroll Swanson Company located in Des Moines. He stressed to all of his correspondents that specially formulated swine rations with antibiotics for each stage of the life cycle were the best way to produce gains for the least possible cost per pound.[10]

Farmers who wanted either to increase hog production or to farrow and market hogs throughout the year were the first people interested in early weaning. In 1954 Lester Heimstra of Cherokee County weaned five hundred pigs at four weeks of age. At weaning time Heimstra separated his pigs into pens of twelve or fourteen animals each and kept them on a prestarter feed for ten days until they were ready for starter feed. At eight weeks the pigs were ready for the growing-fattening ration they would consume for three months until they were ready for market. Farmers such as Heimstra vouched for early weaning, but just like the college experts, they urged farmers who were interested in trying it to practice good sanitation and to provide an adequate supply of clean water. Most farmers were not interested in weaning as early as the experts recommended, however. One farmer stated that he believed in "letting pigs get their milk nature's way for six weeks." A 1955 poll indicated that only 7.3 percent of farmers weaned pigs at four weeks or earlier, while 89.5 percent weaned at six weeks or later. These numbers indicate that in this case farmers were not on the leading edge of innovation. It does show that a large minority (42%) weaned sooner than eight weeks, which was an acceleration of the traditional weaning time. Farmers who tried early weaning found it more profitable than waiting. A Lee County farmer asserted that the pigs weaned at three weeks gained better than the ones he weaned at six weeks of age. This trend in weaning would not have been possible without the presence of growth-enhancing and disease-fighting antibiotics.[11]

As antibiotics became a part of husbandry techniques, manufacturers began to market stilbestrol. The lag between the first publicity of DES in the mid-1940s and the actual introduction in 1954 lasted so long because of time needed to obtain test results on the effects of hormone feeding on the animals. The FDA wanted to ensure there was no danger to meat consumers. It was difficult to obtain consistent results with hormones in laboratories and on experimental farms, which led one author to remark in 1952 that "we can't expect better results under farm conditions and in the hands of non-specialists." While there was the potential for danger to consumers, the consensus among scientists was that only the consumption of abnormally large quantities of meat, especially liver, posed any risk. In 1954, over a decade after the first publicity about DES, the FDA authorized the commercial manufacture and sale of the hormone.[12]

The Iowa State College Cooperative Extension Service released a pamphlet about stilbestrol as a supplement for beef cattle in November 1954. The authors—Wise Burroughs, C. C. Culbertson of the Iowa Agricultural Experiment Station, and William Zmolek, extension animal husbandry specialist—described the benefits of stilbestrol and provided guidelines for its proper use. The authors informed farmers they could reduce feed costs by 10–20 percent, or by approximately two or four cents per pound of live weight. There were also several caveats emphasized in italics throughout the document. Stilbestrol was not to be fed to dairy cattle or breeding animals. Feeding higher than recommended levels was not recommended, and mishandled stilbestrol could pose hazards to humans. The authors emphasized that the bottom line for farmers considering the use of stilbestrol was the difference between costs and return on investment. Feed supplements containing stilbestrol cost from five to ten dollars more per ton than supplements without stilbestrol. Depending on the original cost, farmers could expect a return of between ten and nineteen dollars for every dollar spent on stilbestrol.[13]

A large number of farmers tried stilbestrol almost immediately after it was introduced, and most of them liked it. Forty days after stilbestrol was on the market, approximately 20 percent of Iowa cattle feeders were using it. By the summer of 1955, feeds prepared with stilbestrol were used by 28 percent of cattle feeders. Ample positive publicity helped ease the way for farmers to use the hormone. 4-H leaders sponsored contests for young people to see who could raise cattle with the fastest gains. Farmers who used it gave favorable reports, although farm record keeping was not up to the standards of the records at experiment stations or laboratories. Max Clowes of Humboldt County claimed, "I think stilbestrol did me some good, but I can't prove it. I handled cattle a little differently this year—didn't use as much pasture." Without a control group, Clowes could not make conclusive claims about his experiences. Other farmers such as Ed Hibbs of Pocahontas County based his judgments on his records. Hibbs's steers gained three pounds per day over 163 days with stilbestrol. He stopped feeding stilbestrol thirty days before marketing because he had heard a rumor that stilbestrol-fed cattle did not dress out at the packing plant as well as cattle fed traditional rations. He regretted this decision, however, since the rate of gain for his cattle slowed in those last thirty days. Practical experience with stilbestrol convinced farmers that it was worth not only trying but incorporating into their feed program.[14]

Nine years after the introduction of stilbestrol, Zmolek and Burroughs collaborated to produce an updated guide, incorporating years of continued research and reports from farmers. The most significant difference between the 1954 bulletin and the 1963 version was the degree of complexity, most readily apparent in the title of the new publication, "62 Questions about Stilbestrol Answered." Zmolek and Burroughs repeated some of the basic information, but the new version included less cautionary language and no italicized warnings. The authors reminded farmers not to feed stilbestrol to breeding animals and mentioned that the only residue was found in the visceral organs, not in the meat or fat. As long as farmers observed a forty-eight-hour withdrawal period before slaughter, the trace amount of DES in the liver and other organs was not even detectable. The "62 Questions" bulletin reflected experts' and farmers' optimism about the benefits of feeding stilbestrol.[15]

Feed Additive Fights and Regulation

In the late 1950s several incidents signaled trouble for farmers who used feeds containing antibiotics and stilbestrol. The first was in September 1958 when President Eisenhower signed into law the Food Additive Amendment to the 1938 Food, Drug, and Cosmetics Act. The 1958 amendment included the so-called Delaney Clause, which stated that no food additive could be considered safe if it was a known carcinogen in humans or animals. Researchers knew that DES was a cancer-causing agent when administered in large doses to laboratory mice, although tests on other animals, including cattle, did not indicate a link between DES and cancer. The FDA concluded that medicated livestock feed constituted a food additive, which put stilbestrol under FDA regulation. Stilbestrol remained part of the feeding program for many Iowa cattle producers even though its status as a carcinogen put farmers who relied on it in a precarious position.[16]

Months after the enactment of the Delaney Clause and the FDA ruling on medicated feeds, farmers learned of another warning concerning agricultural pharmaceuticals. This time the warning was from agricultural experts rather than from congressmen. In 1959 the Animal Health Institute and the American Feed Manufacturers issued a joint statement regarding the use of commercial medicated feeds. These groups recognized that such feeds enabled farmers to realize significant increases in production at reduced costs, which made farm products more affordable for consumers. However, "these [medicated] feeds never were designed to be a substitute for sound management and sanitation." According to the two groups, everyone involved in the livestock business—from manufacturers, dealers, and salesmen to farmers—needed to understand that good management was an essential part of the success of medicated feeds.[17]

In the fall of 1959 there was more bad news for farmers, which highlighted the expanded role of the federal government. Just before Thanksgiving, Secretary of Health, Education, and Welfare Arthur Flemming urged consumers to avoid cranberries because of the presence of amino triazole, a herbicide (and carcinogen), on the harvested fruit. In the aftermath of the "cranberry incident," the editors of *Wallaces Farmer* noted that the federal government was prepared to "clamp down" on farm chemicals. According to the newspaper's writers, the FDA planned to increase efforts to enforce the ban on antibiotic residues in milk. Farmers, they suggested, faced a more adversarial relationship with the federal government over the issue of farm chemicals.[18]

Farmers also learned that the chemicals they depended on had problems. By the 1950s it appeared that antibiotics were becoming less effective in promoting growth in livestock than they had been. Over the course of the 1950s farmers observed that flies surviving DDT attacks were breeding new generations that were less susceptible to DDT and other chlorinated hydrocarbon insecticides. Now they learned that their wonder drugs were also becoming less effective. Catron suggested several explanations for this development, including the possibility that there was an increase in populations of resistant bacteria or of organisms that had mutated and were no longer harmed by antibiotics. Catron recommended that farmers use different combinations of antibiotics to prevent the decline in effectiveness. The combination of broad-spectrum antibiotics and good sanitation practices was the experts' ideal.[19]

After the furor of 1959 no news about feed additives was good news for Iowa farmers. The reprieve from government action let farmers conduct their business in ways that suited themselves. They relied on feed with antibiotics for hog production and on stilbestrol for beef production. A minority of hog farmers experimented with new strategies in raising hogs that made subtherapeutic antibiotics even more important. One of the new strategies was to confine hogs on concrete lots with specified rations rather than turning them out on pasture and supplementing their diet with corn. Greater numbers of animals in close quarters created an ideal environment for disease, but antibiotics in feed helped to prevent outbreaks of disease and accelerated weight gain. By the 1960s farmers confined hogs inside buildings to minimize environmental stresses such as temperature change. Some farmers even raised hogs indoors from birth to marketing, a practice known as "life cycle housing" from farrowing to finishing. Because of the changing nature of hog production, feed additives had become more important than ever by the mid-1960s, even as policy makers and the public questioned their use.[20]

The quiet over the issue of feed additives did not last for long. By the late 1960s feed additives were in the news again. On April 1, 1968, the FDA announced that several popular antibiotics would be restricted for livestock production within sixty days if there were no "sufficient objections" raised during that time period. The list of potentially restricted drugs included streptomycin, selected forms of penicillin, chlortetracycline, and Bacitracin. A writer for *Wallaces Farmer* asserted that although the use of antibiotics as therapeutics and as feed additives were concerns to the FDA, it was their subtherapeutic use in growth promotion that was the most problematic. The Iowa Pork Producers Association and the National Pork Producers Council based in Des Moines objected to any FDA restrictions. Rolland "Pig" Paul, president of the National Pork Producers, wrote a letter to the FDA urging a delay on any ban. Paul wanted to know if the issue of antibiotic residue in meat products was significant enough to warrant a ban. He assured FDA officials that hog farmers wanted consumers to have safe meat and asserted that the degree to which those residues actually constituted a health threat was not clear to producers. Similarly, a representative of the Iowa Pork Producers stated that safe meat was "essential for the survival of the [livestock] industry," but that any regulations had to be "workable." The definition of "workable" for the hog farmers included maintaining antibiotics for therapy and for feed, which was precisely the problem for the FDA scientists.[21]

The issue of chemical residues in meats resurfaced in 1970 when the USDA, charged with inspection of meatpacking plants and meat grading, became part of the system of reporting and oversight for chemical residue. An anonymous official at an Iowa meatpacking plant reported he had a list of ten Iowa farmers who were out of compliance with acceptable levels of residues in animals they had marketed over the previous several weeks. "The next time these people market livestock," he stated, "the federal meat inspection division in Washington wants tissue samples from their animals to check for drug and pesticide residues." The federal government was prepared to hold individual farmers accountable for the ways in which they used chemicals.[22]

Officials at the USDA and the office of the Iowa state veterinarian explained the new compliance program. Animals that tested above the tolerance level of chemical residues in tissue samples would be withheld from slaughter or destroyed. If the samples showed residues in a range from the official tolerance to 20 percent of that level, animals from the farm of origin would be sampled the next time the farmer marketed livestock. If liver tissue samples from a packing plant indicated that drug residue levels were at 20 percent of the tolerance level, the lab would notify the packing plant of origin that they were at risk of being cited for violations. This approach assured that farmers would have some warning if they were at risk of violation, and it would bolster consumer confidence that animals with significant levels of chemical residues would not enter the food supply. In spite of these measures, however, the director of the USDA program stated there were several shipments of livestock from Iowa that exceeded established tolerance levels for residues.[23]

Some cattlemen simply did not follow the guidelines for chemical use, notably stilbestrol. In 1958 experts recommended withdrawing stilbestrol from feed forty-eight hours before marketing. This allowed DES to clear the animal's system before slaughter. But when the new USDA compliance program began in 1970, cattle arrived at packinghouses with DES levels that exceeded tolerance. The state veterinarian reported that over the previous year he had contacted "about a dozen" farmers regarding residues. Furthermore, feed additives were just one category of chemicals showing up in tissue samples. Chlorinated hydrocarbon insecticides were also present, although stilbestrol was the most common drug found in cattle tissue samples.[24]

Iowa State experts confirmed what packers, the FDA, and USDA already knew: a sizable minority of farmers did not understand feeding guidelines. An Iowa State study from 1966 indicated that among farmers who raised beef cattle 79 percent recognized that stilbestrol should be withdrawn from cattle rations forty-eight hours before marketing, while 12 percent of farmers disagreed with the forty-eight-hour statement. It is unclear if they believed the actual rules were more or less stringent. The fact that a remaining 9 percent of cattle producers had no opinion on the withdrawal rule was a serious problem, for it suggested that a significant minority of farmers did not understand the rule, could not understand it, or ignored it.[25]

In 1971 there were new efforts to control the farmers who misused stilbestrol. In January the American National Cattlemen's Association and the National Livestock Feeders Association announced a new voluntary program to prevent DES contamination in the meat supply. Beginning in March, feeders would sign certificates to assure packers that the cattle being marketed had not been fed DES for at least forty-eight hours prior to slaughter. Feed manufacturers and organizations such as the Animal Health Institute supported the program and offered pamphlets outlining the new program to farmers, farm journalists, and members of farm organizations such as the Future Farmers of America.[26]

The certification program did not succeed in changing the ways farmers used stilbestrol. In the summer of 1971 a poll of Iowa farmers indicated that 48 percent of them believed their neighbors did not follow the recommended withdrawal period for DES, and 44 percent of farmers reported they were unsure whether their neighbors were in compliance. The most damning figure, however, was that 8 percent of farmers knew at least one person who did not observe the prescribed withdrawal period. If even a handful of producers failed to follow the guidelines, then those producers could jeopardize the future of DES and other medications.[27]

Most farmers who wanted to use drugs and medicated feeds recognized the risks of noncompliance. According to Don Lefebure, a hog farmer from Linn County, "It's better and easier to follow withdrawal recommendations than to have the FDA or some other government agency enforcing stringent regulations." Lefebure was not alone. "I don't agree with some of the withdrawal requirements," echoed an unidentified cattleman, but then he continued, "If they [government inspectors] found residues in the meat I sold, I'd be out of business." These farmers urged their peers to use medicated feed properly in order to ensure that the drugs would remain available for all producers. Roy Keppy, president of the National Pork Producers Council, explained that farmers were meat consumers too. "I don't want anything in it [meat] that's detrimental to my family or other consumers," Keppy noted. Farmers confronted a dilemma. They used drugs to reduce production costs and to satisfy consumer demand for low-cost meat, but the continued use of those drugs then potentially alienated consumers and their government representatives in Congress and regulatory agencies.[28]

In 1971 the USDA and FDA changed the stilbestrol withdrawal period from forty-eight hours to seven days. Inspectors found that ten out of approximately twenty-five hundred liver samples from slaughtered cattle indicated the presence of DES. Pressured by the public and congressmen to ban stilbestrol, government agencies could not afford to be idle. Regulators wanted to prevent a ban by lengthening the withdrawal period. They hoped that recalcitrant producers would see the importance of compliance and voluntarily change their behavior. FDA and USDA representatives justified the new seven-day withdrawal period, stating that "we simply cannot have any DES residues in the food supply" and that producers shared a large part of the responsibility for reassuring consumers that the meat supply was safe. On January 8, 1972, regulators made the certification program mandatory.[29]

Farmers' experiences with the certification program were not encouraging, however. In early 1972, one year after the voluntary certification program began and several months after the commencement of the mandatory program, a majority of farmers surveyed stated that livestock buyers had not asked them to sign any sort of drug compliance statement. Only 18 percent of farmers stated that buyers had asked them to sign. The good news for farmers was that feeders raising the largest numbers of animals were the more likely to sign a withdrawal form. Buyers asked 55 percent of farmers who fed more than two hundred cattle to sign and 30 percent of hog farmers who raised more than five hundred head. The withdrawal statement program failed to accomplish its goals. As long as 45 percent of large-scale beef producers and 70 percent of large-scale hog producers did not sign drug statements, it was likely that many animals would arrive at the packing plants with drug residue in their bodies.[30]

Farmers and their representatives in the U.S. Congress fought for the right to continue using feed additives. In summer 1972 there were several bills pending in Congress to ban DES. In August the FDA announced it would ban stilbestrol in livestock feed after January 8, 1973. Farmers could still use DES implants placed behind the ear of cattle. Two members of Iowa's congressional delegation cosponsored a bill to allow the continued use of stilbestrol. They argued that the only way to get cancer from DES was to ingest massive quantities of it. If the FDA was not bound by the Delaney amendment, they argued, the FDA would be able to guarantee a safe food supply without banning the drug. This effort was unsuccessful, however. The ban on stilbestrol as a feed additive took effect on January 8, 1973.[31]

Farmers pledged to continue using DES after the ban, even though they could not use it as a feed additive. Approximately six hundred cattle feeders met in Ames in July 1973 and signed a petition urging the federal government to allow "realistic" amounts of DES in cattle tissues at time of slaughter. They argued that trace amounts were safe and that a total ban would increase farm costs so much that it would be difficult to stay in business. Implanting DES required more care and labor than feeding, since it was a separate task and numerous implanting errors were possible. Common implanting errors included gouging ear cartilage, severing ear veins, and implanting the pellet between layers of skin instead of under the skin, which prevented proper absorption of the hormone. Robert Lewis of Mitchell County estimated that a switch to implants would increase production costs by as much as 15 percent for his herd of eleven hundred cattle. The more serious critique of the stilbestrol ban in feed was that implants were not removed before slaughter. The implant continued to release the hormone up to the time of slaughter, leaving farmers in the same position as they were when some farmers continued to feed stilbestrol-laced feeds beyond the specified withdrawal period. The executive vice president of the Iowa Beef Producers Association stated that the ban was less about a safe food supply and "more of a psychological and emotional action than anything else."[32]

Meanwhile hog producers were also facing increased restrictions on feed additives. In January 1972 the FDA recommended banning several antibiotics that were popular in hog production and tightening controls on other antibiotics. At a pork producers' meeting, ISU swine nutritionist Vaughn Speer explained that FDA officials feared the use of antibiotics in livestock feed could potentially cause resistance in microorganisms harmful to humans. The swine editor for *Successful Farming* speculated that bans or limits on antibiotics could be even more serious for hog producers than a ban on DES would be for beef producers. As farmers raised larger herds and used more confinement systems, antibiotics were "the crutch" that kept them in business. The modern hog business would collapse without antibiotics.[33]

The presence of other chemical residues in animal tissues contributed to the problem facing farmers who wanted to preserve those chemicals as an important part of livestock husbandry. An Iowa hog farmer who used an arsenic compound to control an outbreak of swine dysentery in his herd accidentally overdosed one group of hogs in his herd. The USDA veterinarian took tissue samples from twenty-one of the fifty-two animals, and all the carcasses were frozen during the analysis. Test results showed that the levels of arsenic were within acceptable tolerances, and so the meat entered the food supply, although the heads and viscera were condemned. The practical lesson farmers learned from this and similar incidents was that it was costly and time-consuming to conduct the tests, freeze the carcasses, and then have part of the by-product destroyed.[34]

As the 1970s began, farmers were on the defensive. In the 1950s they had been eager to obtain the antibiotic and hormone wonder drugs. Farmers used antibiotics and hormones to speed the rate of gain and to accomplish those gains with less feed. Feeding chemicals was a survival strategy for livestock producers, but it was a troubling strategy. People far removed from agricultural production debated the merits of livestock-husbandry practices that boosted production yet posed unknown risks. Farmers believed the decisions about which drugs they could or could not use and the withdrawal periods for those drugs were based on fear, not fact. They in turn behaved without fear and fed stilbestrol in violation of the guidelines from manufacturers, experts at Iowa State, and government regulators. Consequently they lost the fight over stilbestrol in cattle feed. The fight over antibiotic feeding did not end in the 1970s. It was unclear how much, if any, antibiotic residue would be acceptable in hog tissues. Faced with concern from prominent members of Congress, advocate-journalists, and members of the public, farmers defended themselves and their production choices, just as they defended the other decisions they made about pesticides and fertilizer.

Part Two   MACHINES

Five

Push-button Farming

The farmer lay dozing in bed as his alarm went off on a cold February day. It was three o'clock in the morning. Instead of rising, pulling on several layers of clothes, boots, and overshoes and heading for the barn to prepare for the morning milking, the farmer stayed in bed. He simply stretched out his leg to a console on the bedroom wall with buttons marked "Chores," "Feeding Cows," "Slopping Hogs," and pushed the button marked "Milking" with his toe. There was no rush to get to work. A sign on the foot of the bed, saying "In Conference until 12:00 Noon," indicated that the farmer was in no hurry to get to work. This imaginary scene was the creation of a cartoonist for *Wallaces Farmer* in January 1950. The title, "Pushbuttons by '60?" suggests this was a humorous take on a serious issue: the role of mechanization on Iowa farms in the postwar period. Specifically, which of the time-consuming jobs such as milking, hauling milk in cans, shoveling manure, and scooping feed for livestock would be mechanized in the years to come? To what extent would farmers reduce the amount of physical labor required to operate a farm? How would farmers use push-button techniques to change the ways they farmed?[1]

The idea of automated or push-button technology was a graphic one for farmers who did the milking, feeding, and manure removal chores 365 days of the year. Labor-saving devices have always appealed to some farmers, but the drive for labor savings accelerated in the postwar years. Push-button imagery suggested a modern world of comparative ease, one in which farmers could leave behind the stoop labor of agriculture. By the 1960s the rhetoric of push-button farming gave way to the language of automated materials-handling systems, but the appeal to farmers was the same. Automated materials handling replaced the scoop shovel for many jobs on Iowa farms in the postwar years, although the degree to which that change occurred varied from farm to farm, depending on the degree of specialization and size of operation.

Push-button techniques represented the ultimate fulfillment of the industrial ideal in farming. Unlike many of the harvest tasks that farmers mechanized, dairying, feeding livestock, and moving manure were everyday chores that on most farms took several hours a day. Automated materials-handling systems could be used to specialize or expand production. Farmers with small acreages or small herds as well as those with larger operations benefited from the economies made possible by push-button techniques. Dairy farmers were among the first to use these techniques, installing new milk-handling systems in the 1950s, in response to various factors: the widespread availability of electricity, rising labor costs, and more stringent sanitary requirements for producers of Grade A milk. Families feeding beef cattle were quick to adopt automated feeding systems, although the scale of automated operations varied depending on the size of the operation. Finally, farmers used automated materials handling to move beyond the feedlot. They concentrated large numbers of animals together in confined spaces, sometimes indoors. Automated systems fed the animals, watered them, and even removed manure. Confinement feeding of large numbers of hogs and cattle became a possibility during the 1960s, though as the decade ended, farmers who practiced confinement feeding were dealing with a growing waste-management problem.

The combination of the shortage of farm labor and the spread of rural electrification made push-button techniques an option for farmers in the 1950s. In Iowa rural electrification was just underway by the time World War II began. Approximately 34 percent of the 209,737 occupied farmhouses in the state claimed electric power in 1940. During the decade of the 1940s, however, the electrification of farmsteads proceeded quickly, and by 1951 almost all Iowa farms had electricity. Farmers of the 1930s could only envision installing a few electric lights in their homes, barns, and outbuildings and powering a few appliances at a cost of approximately twenty-five kilowatt hours per month. By 1950 that situation had changed for a growing minority of farmers, who now used over one thousand kilowatt hours or more of electricity per year for many farm tasks. Farmers realistically conceived of using the new technology to do much of the repetitive and strenuous stoop work on their farms at a time when there were fewer strong backs to do the work and those laborers who remained demanded higher wages. With the decline in farm labor and the rise in rural electrification, an electricity-powered materials-handling system became the new "hired man."[2]

Rebuilding the Dairy

Dairying was—and remains—one of the most labor-intensive of all tasks and work cycles on the farm. Cows need to be milked twice a day, barns need to be cleaned every day, milk must be stored until it can be hauled to a processor. The tools and implements that come into contact with milk require cleaning and sanitizing before they can be used again. In the late 1940s several developments reduced some of this labor and attracted attention: the pipeline milker, the bulk tank, and the milking parlor. With a pipeline milker, the milk flowed from the automatic milkers directly from the cow into a glass container, then through tubes into a bulk cooling tank. In 1950 hand milking was still the norm for over half of all cows in Iowa; only 45 percent of cows were milked by machine. Both techniques, however, exposed the milk to contaminants, including dirt, manure, insects, and airborne pathogens. The person who did the milking also carried and poured the milk into cans, then moved the cans to the milk house for cooling in a water bath. Taken altogether this was a staggering amount of labor. In 1951 Gerald Prince of Guthrie County calculated that he had carried heavy milk cans over 88 miles and walked a total of 375 miles in his dairy operation during the previous year. Pipeline milking systems eliminated the need for this physical work. The milk moved by machine and freed the farmer to inspect cows, wash udders, and operate the machines. Milking parlors were built with raised stalls for the cows so the farmer could clean and inspect udders and attach the milking machinery without bending over. These developments eased the burden of dairy farming considerably.[3]

The most significant appeal of new technology was the ability to increase the pace of work and reduce labor costs. With labor expenses accounting for approximately 25 percent of the cost of dairy farming, any reductions increased profits. When a family from rural Kalona used a milking parlor and bulk system to reduce costs of transporting milk to Iowa City, the reduction was from thirty-five cents per hundredweight to twenty-two cents. The layout of new milking parlors varied, but one of the popular styles after 1957 was the herringbone type. This type of setup resembles car parking on a one-way street of a small town. Just as cars face the sidewalk at an angle with their taillights to the street, cows in a herringbone milking parlor face their feeder with their tail ends angled toward the operator's pit. A gate closed behind the last cow and the close quarters kept the cows in place. A pull on a rope or chain released a fixed

amount of feed from overhead storage bins into the feed cups. The operator could see the udders and attached the pipeline milkers to the cows as
they closed in. Pipeline milkers carried the milk to a bulk tank, which
eliminated the need to move the milk by hand from each cow to a can,
then the can to a milk house where it would be placed in cool moving water. The new system allowed farmers to milk much more rapidly. As soon
as one group of cows was finished, another group was already queued up
outside the door ready to be milked. The old way was to take an individual milking unit to each cow in a stanchion.[4]

Dairy farmers commented favorably on the increased pace of milking
in the new-style parlors. Kenneth Showalter of Franklin County claimed
that he could milk his 23 cows by himself in one hour with his new milking parlor. Formerly Showalter and his hired man had each spent an hour
milking with a stanchion system and portable milking machines. "This
new set-up saves from two to two and a half man-hours per day," Showalter concluded. In the mid-1950s Rudolph Remmen of Winneshiek County
decided to upgrade his facilities. His stanchion barn only had room for 23
cows, so he constructed a new milking parlor to accommodate 12 cows at
a time and expanded his herd to 52 cows. With an average milking time
of fifteen minutes for each group of 12 cows, Remmen was able to milk
the entire herd in approximately one hour. Farmers with larger operations realized larger labor savings. The Hermanson brothers of Story
County kept a herd of 150 Holstein cows and were one of the first operations in Iowa to use the herringbone-style parlor. They could milk between 50 and 60 cows per hour in their double-six herringbone. According to Leonard Hermanson, "It took three men all day just to milk and
feed the cows [in stanchions]. Now, two men can handle the same work
in about six hours."[5]

Creature comforts were important benefits to the new-style milking
parlors. They were easier on the operator than the old stanchion milking
barns in which farmers had to constantly bend over to clean udders,
check for infections, and attach or remove milking machines. New milking parlors were self-contained, either as separate buildings or enclosed
spaces within barns, making it easier to control the environment by keeping the dust down, keeping flies out, and even providing heat in the winter. Three Wright County boys claimed to enjoy milking in their new
milking parlor. They installed a radio and a propane heater to make the
work more pleasant. Vernon Miller of Black Hawk County stated that the
raised stalls made it easier for his father, who suffered from rheumatism in
his knees, to assist with milking. Eldon Johnson, also of Black Hawk
County, commented that working in the pit was comfortable. "And it
eliminates the danger of being kicked. Also, the udder is in clear view and
easy to wash." While farmers did not cite comfort as the most important
benefit of milking parlors, they certainly enjoyed the new ease of work.[6]

Bulk tanks were a critical part of the new style of dairying. These electric-powered cooling tanks held several hundred gallons of fluid milk. Bulk-tank technology was initially developed in California, where some of the first large-scale dairy operations began. Farmers could either pour the milk from the milking units into the refrigerated tank by hand or they could use pipeline milking machines to transfer the milk automatically from cow to tank. Once the milk was in the tank, the farmer had only to wait for a dairy tank truck to pick up the milk on an every-other-day basis. In addition to labor savings and reduced physical labor, there were other advantages of bulk handling as well. Less handling meant less opportunity for spillage, and rapid cooling meant less bacteria growth. Promoters of bulk handling suggested also that dairy work was more attractive to hired labor because of the reduced physical labor requirements.[7]

Bulk handling allowed farmers also to produce fluid milk that would be classified as Grade A, in other words, milk for drinking. New standards for farmers who sold fluid milk posed a major challenge to farmers who wanted to produce for the market with the highest prices. The Iowa legislature passed a law in 1951 that enacted more stringent sanitation requirements for processors marketing Grade A product. The milk at Grade A processing plants could not test in excess of a bacteria count of 400,000 per million, and finished milk could not exceed 30,000 per million. The E. coli count of finished products could not exceed 10 parts per million. If farmers could not provide raw milk from disease-free herds that met these standards, they were limited to producing for the less profitable Grade B market. Over the course of the 1950s, concerns about making the grade in milk production was an important reason for farmers to consider the new dairy technology.[8]

The potential for expansion was an important consideration for farmers studying the merits of bulk handling. A farmer who owned a cooler for milk cans claimed that his can cooler was too small for his production and a bulk tank was a good investment. Farmers who owned bulk tanks wanted to keep them full, which helped them pay for the system as soon as possible. Alex Young, a North Liberty farmer who produced for the Iowa City market, used his pipeline milker and 400-gallon bulk tank to increase his herd from fifty cows to seventy cows without increasing his workload. Young stated that "the only thing left for me to do is to put the milker on the cows. Of course, I have to keep the equipment clean. But the tank is easier to wash than milk cans, and the pipe doesn't take a great deal of cleaning time." In 1958 a Benton County farmer noted that his two-year-old 250-gallon tank was not big enough to handle production from his expanding herd. He planned to trade in for a larger model. In the interim he asked the local dairy processor to come every day to pick up milk instead of every other day.[9]

Pipeline milking and bulk tanks reduced labor and allowed farmers to expand, but automatic feed grinding and moving feed to the cows also reduced labor requirements. Scooping each pound of feed to cows was a production bottleneck that limited the profit potential of the new technology. Dairy farmers frequently mixed supplements with grain or silage to provide a more balanced ration for lactating cows, and this was time-consuming work. Norman Amundson of Clayton County ground, mixed, and even bagged his own feed before feeding it to his cows by hand, although in the 1950s he had his feed ground and mixed in town and delivered to his farm. He installed a hundred-bushel steel feed bin and an electric-powered auger system to deliver the feed to a device that metered out feed to each cow in the parlor.[10]

The majority of feeding for milk cows, however, took place outside the milking parlor. Farmers traditionally grazed milk cows on pasture in warm weather, and during the colder months they fed forage and silage that comprised the bulk of the dairy ration in outdoor lots or enclosures. In the 1950s, due in large part to changes in forage harvesting, farmers began to feed cows in lots throughout the year. This was called drylot dairying. Mechanized forage choppers cut and chopped the crop into short lengths, and it was then blown into a forage wagon equipped with an automatic unloader. By feeding cows in the lot instead of grazing them, farmers maximized production from their forage acres. A Wisconsin study indicated that ten acres would support seven cows with rotational grazing while those ten acres could support fourteen cows if the grass was harvested by machine. This type of dairying meant more hauling of feed but also more production per acre of grassland.[11]

New dairying was considerably more expensive than pasture grazing and stanchion-barn milking. Costs varied depending on the size of the parlor and the milk-handling system, but farmers could expect to pay at least $6,000 for a parlor to accommodate eight cows at a time during the 1950s. Costs were lower for milking systems suited for larger herds. According to a Michigan State University study, a herringbone system designed for forty-five cows an hour could cost as little as $6,040 while a system for twenty or thirty cows an hour was estimated to cost $6,365. Similarly, a University of Minnesota study indicated that bulk tanks were more economical for larger producers than for smaller producers. Bulk tanks for Iowa producers cost from $2,000 to $3,000, depending on tank capacity. Farmers who produced 250 pounds of milk per day would expect to pay twenty-eight cents per hundredweight for a 200-gallon tank. A farmer who produced 650 pounds of milk per day paid eleven cents for the same-sized tank. These costs included depreciation, along with estimated taxes, repairs, insurance, and interest charges. This was one of the most expensive capital investments a farmer could make outside of land and livestock purchases.[12]

One Delaware County operation shows the degree of capitalization required to set up a modern dairy farm. Howard Stone, his son Jim, and neighbor Don Miller formed a partnership to update a farmstead that included a stanchion barn and loafing shed (a shelter for cattle to get out of the weather) as well as two silos and a milk house for storing milk as it cooled. They designed the new system to feed silage and chopped hay in the feed bunk from a self-unloading wagon and fed cows in the milking parlor from overhead bins. They invested $8,000 in a new milking parlor, milk-handling system, a new loafing shed that doubled the size of the previous structure, a straw-storage shed, and feed bunk.[13]

The high first-cost of new milking parlors, pipeline systems, and bulk tanks posed a dilemma for farm families who sold fluid milk or produced cream for sale and for use at home. Should they continue dairying and expand their herds to realize the economies of scale? Most dairy producers answered in the negative. They chose instead to engage in other farm tasks involving less financial risk than dairying. Throughout the 1950s and 1960s the number of farms dedicated to dairying dropped quickly while the average number of animals per dairy herd increased. Farmers who maintained a small herd of approximately half a dozen cows to produce for home dairy needs and to sell some cream had no use for the new dairy equipment. Those who did not invest in the pipeline milkers, bulk tanks, and milking parlors invariably left dairying or farming altogether. A 1959 survey of Iowa farmers indicated that the new dairy techniques were the tools of the minority rather than the majority. Only 12.5 percent of respondents owned bulk tanks, 11 percent owned pipeline milkers, and only 6 percent had milking parlors.[14]

Expansion was a source of tension for dairy farmers. Alvin Brown of Black Hawk County declared, "A man is going to have to dairy in a big way—or not at all." Rather than invest in the new equipment, Brown sold his herd of twenty cows in the winter of 1954–1955. Another farmer, who increased his herd, noted, "It's to the point now where you either get in or get out." When Hardin County farmer Roy Engel's milk buyer began to take bulk milk, Engel sold his herd of fourteen cows, explaining, "I never enjoyed dairying too much anyway, and I didn't want to get in so deep that I couldn't get out if I wanted to." Engel sold milk as part of a diversified operation, as a hedge against crop failure or price declines for crops or livestock. But in the new world of dairying, he found that continuing to produce milk was not worth the expense of expansion. In 1961 Harry Clampitt, a Hardin County farmer and the president of the American Dairy Association of Iowa, expressed his view of contemporary changes in dairying. He questioned "whether going into dairying on a 10–15 cow size is going to be popular or profitable." As the 1960s began, an increasing number of farmers ceased dairying and those who remained increased their scale of operations. Attrition out of dairy farming meant that the technological solutions of the minority became commonplace on dairy farms.[15]

Farmers who chose to remain in dairy production milked more cows. The cost of the new equipment compelled farmers to maximize production in order to pay for it. Melvin Hoelscher of Hardin County was one of those farmers who expanded, after he invested over three thousand dollars in a new bulk tank and pipeline system. "I've got to get [the investment] back somehow. And the only way to do it is to milk more cows." A northeast Iowa dairyman expanded his herd dramatically. In the early 1950s he milked a herd of between 15 and 18 cows per year. By 1955 he had expanded to 37 cows and hoped to milk from 40 to 50 head. "I was ready to quit," he noted. "But the wife and boy said they'd do the milking if I'd build 'em a new milking parlor." John and Fay Hostert of Dubuque County increased their herd from 50 to 70 cows in 1957. They reasoned that it was worth expanding because they already owned the new equipment. "And," they explained, "it's no more work to milk 70 cows than it was 50." These farmers found that technology solved their short-term problems of high labor costs and of gaining access to the higher-priced Grade A fluid-milk market. Expansion with new technology was the only way farmers believed they could increase profits and remain in dairying.[16]

The number of dairy cows and production figures per cow reveal the scope of the change wrought by farmers using the new technology. In January 1951, the year that state requirements for Grade A milk production became more stringent, there were 1,007,444 cows and heifers for dairy production on Iowa farms. By the beginning of 1962 there were only 769,810 dairy cattle in the state, yet they produced more milk than ever before. There were also fewer farmers in the dairy business. The number of farmers with herds of 20 or more cows increased from 1950 to 1955 by one-third. With the new equipment, farmers were able to increase their production and expand their herds.[17]

The experiences of dairy farmers in the Sioux City milk shed demonstrate the consolidation. The number of Iowa, Nebraska, and South Dakota farms that produced fluid milk for the Sioux City market declined precipitously from 1950 to 1960. In 1950 there were 730 dairy producers, in 1955 there were 480, and in 1960 there were just 170 producers in the milk shed. However, the production for that market increased, which more than offset the loss of dairy farmers. In 1955 Sioux City processors handled 34 million pounds of milk from 480 producers. In 1960 they handled 76 million pounds of milk from only 170 producers. This concentration of dairy production and increased productivity was testimony to the efficacy of new techniques adopted by Iowa dairy farmers and their families.[18]

Two other developments became popular on Iowa dairy farms as expansion and consolidation prevailed—loose housing and free-stall housing. The old-style stanchion barns had designated positions for each cow for feeding and milking, which held cows in their place during milking. In cold weather cows often remained in the stanchions. By contrast, a loose-housing barn was simply a structure without the rigid organization of individual stanchions for each cow. Instead of milking in the barn, farmers milked the cows in the specialized parlors described previously. Cows moved around the loose-housing barn at will. The floor was packed earth covered with straw and animal manure, which generated heat in the winter to keep cows warm. Cows ate from a common feed bunk in the outdoor lot and another one in the milking parlor.[19]

Loose-housing systems were as distinct as the individual farmers, reflecting tenure status and available capital. Landowners typically had more extensive livestock operations, since the buildings, lots, and equipment remained on their own property. J. Clifford Grant, a farm owner from eastern Iowa, invested thousands of dollars in his system. In 1950 Grant planned to expand his herd and wanted to construct a stanchion barn for seventy-two cows. The contractor who bid on the project encouraged Grant to consider loose housing. After viewing existing loose-housing operations and consulting with experts, he accepted their advice and built several new truss-framed structures to make mechanized cleaning with a tractor and loader easier than when dodging support poles. He added a steel silo and a holding shed, too. Grant's new operation was expensive, although it was possible to spend less on loose-housing systems. Donald Glew, a Delaware County tenant farmer, converted a stanchion barn into a loose-housing barn for only $387. He removed the stanchions on one half of the barn for loose housing for twenty-four cows and used the other half for his milking room and storage area.[20]

Loose-housing systems were much less labor-intensive than stanchion milking. Farmers who used pail-style milking machines for a thirty-cow herd in a stanchion barn hauled eighteen tons of material per cow annually. This figure included four tons of milk, two tons of hay, three tons of silage, one ton of straw, one and a half tons of grain, and six and a half tons of manure. This hauling and shoveling amounted to one and a half tons per day. Given the amount of work involved with stanchion milking, it is not surprising that some farmers wanted to change. Kenneth Winkel of Lyon County explained that he studied many different plans for dairy structures before he tore down his old stanchion barn and replaced it with a modern facility. After using it for only one year he claimed, "I would stop milking cows before I would go back to a stanchion system."[21]

Labor savings were a significant feature of loose housing. Instead of cleaning out the gutters behind the cows in stanchions every day, farmers cleaned the loose-housing barn on an occasional basis. In the new system, milking took place in the parlor, which was kept very clean, at least in ideal operations. The cows rested on bedding in the loose-housing barn and were protected from the manure and mud that they may have lain in when either in the pasture or in the stanchion barn. In the loose-housing barn they were on a mat of bedding, which would keep them out of the animal waste. The alleys would be cleaned out periodically. Dick Moen of Howard County scraped the entire surface of the barn once per year.[22]

Farmers cited herd health as an important advantage of loose-housing over stanchion-barn systems. In 1950 Vernon Lyford tore out his stanchion barn and replaced it with a loose-housing setup on his Worth County farm. During his days of milking in the stanchion barn, Lyford's cows had endured a bout of mastitis, an infection of the cow's udder that prevents milk production in one or more quarters of the udder. Since he built the new barn, with cows resting on their own without being confined in the stanchions, there were fewer health problems. "Right now," he stated in 1953, "I'm milking 34 cows—every one of them giving milk out of four quarters [of the udder]." Keneth Winkel claimed that there were numerous advantages of loose housing, including greater comfort for cows, cleaner cows, and fewer bruised teats and infections such as mastitis. Even in loose-housing barns, farmers invariably coped with illness or health concerns that cut production, so providing an environment that was more conducive to good health was the best way to increase production and profit.[23]

Free-stall housing, similar to loose housing except with stalls provided for cattle to use for rest, was also considered a healthier alternative for cows. With individual stalls instead of open space, farmers saved on the amount of bedding. Gerald Ehlinger, a Dubuque County dairyman, constructed a free-stall barn with two rows of four-by-eight-feet stalls on either side of a twelve-foot-wide alley. The floors of the stalls were of packed earth covered with straw and were slightly elevated above the alley floor. Ehlinger claimed that his cows stayed clean all winter long with much less bedding than in a loose-housing system. Farmers experimented with bedding material, using straw, chopped cornstalks, and even sawdust in the stalls. By the late 1960s sawdust had become the material of choice. According to Harold Leazer of Cedar County, "Sawdust alone keeps the cows cleaner and dryer than anything I've tried [before]." Ray Crock Jr., also of Cedar County, raved about the cleanliness of the cows housed in stalls with sawdust. "They're just as clean as during the summer and maybe even a little cleaner." Regardless of the particular bedding material, farmers believed that loose and free-stall housing would provide a cleaner environment for their cows.[24]

Feedlots

Families who specialized in dairying were not the only ones who employed automated materials handling and changes in architecture to reduce labor costs, expand operations, and ease the hard work of farming.

The largest growth in automated materials-handling systems was among farmers raising beef cattle in feedlots, since many more Iowa farm families depended for their livelihood on the sale of beef cattle than on the sale of fluid milk. Just as most Iowa farm families kept a herd of dairy cows for the sale of cream instead of fluid milk, they also kept a herd of beef animals. This production strategy enabled them to minimize the risks of farming by selling a variety of products. If hogs were selling low, then cattle might be selling high. Families who liquidated their dairy herds often increased their commitment to feeding beef cattle. In the atmosphere of postwar prosperity, urban consumers wanted more beef. In contrast to dairying, expansion of beef herds was comparatively inexpensive, since feeder steers were less expensive than breeding stock such as heifers and cows.

Expansion was also labor-intensive, since farmers needed more physical labor to feed those animals. While cattle could be grazed on pasture or in harvested cornfields to feed on cornstalks and any remaining ears of corn, farmers needed to "finish" cattle before they could be sold. Finishing was a period in which the animals would be fed grain, forage, plus any number of supplements to increase their weight rapidly and command a top price. Farmers needed to get feed in long troughs (called feed bunks) at least twice daily. For even a small herd this amounted to tons of feed every season, which had to be moved by hand.

Over the years farmers used a variety of labor-saving devices to help feed livestock. Numerous companies manufactured track systems that could be mounted to the ceilings of barns. Suspended cars could be pushed along the track, allowing farmers to move feed or even manure from place to place without carrying it. Some farmers developed their own systems, making carts that could be moved by hand along the length of a feed bunk, with the farmer shoveling feed out of the cart at intervals along the bunk. In 1948–1949 Leo Boddicker of Benton County installed a carrier system that enabled one man to feed two hundred steers on a 335-acre farm. Almost all of these innovations relied on hand power for loading and for propelling them along the track. There were physical limits to how much one person could move, which consequently limited the number of animals that one person could feed.[25]

In 1954 the magazine *Successful Farming* highlighted the role of automated materials handling in a special issue. The information for this special feature came from a USDA and Illinois agricultural experiment station study on the economies of labor saving on thirty-six Illinois farms. As the authors noted, most farmers of the early 1950s continued to feed forage and grain the same way farmers had done for a generation or more. This made sense for farm families that fattened less than twenty head of cattle. The USDA recommended that hand feeding was still the most profitable system for those who fed herds of less than fifty animals, which described most Iowa farmers.[26]

Hand feeding was too labor-intensive for farmers who hoped to beat the cost-price squeeze and offset shrinking profit margins by producing more. The most common solution for farmers who wanted to stretch their labor over larger herds was to use new self-unloading wagons to feed silage, or grinder-mixers to feed grain and other feedstuff. Farmers pulled these wagons with a tractor alongside feed bunks and the self-unloading wagons deposited feed in the bunks. In this way it was possible to move tons of feed without the farmer having to leave the tractor. The problem was that during the spring thaw it was often impossible to drive through cattle lots.[27]

Cattle feeders mechanized with fence-line feeding systems where the bunks themselves became part of the fence. Driving into a feedlot meant having to contend with gates, or with getting the tractor and load of feed stuck, or with animals crowding around the bunks or escaping when gates were open. Fence-line feeding allowed farmers to avoid all those inconveniences. They simply drove along the fence to unload the feed into the bunks, and the work was done. This technique quickly became a popular method of automated feeding. With self-unloading wagons one person could haul feed for hundreds of animals.

To gain this kind of economy, farmers needed to conduct careful planning for structures, lots, and bunks. Because the bunks were part of the permanent fence, they could not then be moved to avoid muddy areas, like portable bunks. As one Pottawattamie County cattleman pointed out, "We're afraid of mud where the cattle stand and also where you have to drive for unloading." These problems could be overcome with concrete next to the bunks for the cattle to stand on and gravel pathways for the tractor, but these measures increased the cost of fence-line systems.[28]

Self-feeders were a less expensive alternative than fence-line feeding, since a farmer could fill the feeders with a large quantity of grain and let the animals eat for several days before refilling. These devices could be designed for hogs or cattle, with wooden units for cattle and galvanized metal for hogs, since hogs tended to be harder on the feeders than cattle. A typical self-feeder had a central storage unit and feeding areas at the bottom. This allowed gravity to pull feed downward as the animals ate. Farmers used grinder-mixers to prepare their own rations (consisting of grain, supplements, and forage crops) and to deliver the feed directly to bunks or self-feeders. Ten years of experience confirmed the value of self-feeders for Hans Hansen of Emmet County. He liked the fact that during busy periods he was not tied down to a time-consuming regimen of livestock chores.[29]

Farm families who did not own their own grinding and mixing equipment could have bulk feed delivered directly to their self-feeders. This allowed farmers with small herds to avoid the expense of investing in equipment and also allowed them to spend their time on other farm jobs, especially during corn planting, haymaking, and forage making, and corn and soybean harvesting. As Merrill Randau of Story County explained, "Home grinding ties up both a tractor and a man. I couldn't afford to do my own grinding, especially when field work is pressing." Claude Ruckman of Hardin County praised the labor-saving qualities of custom-filled self-feeders. "I take my corn to the elevator and they deliver the mixed ration directly to the feeder. All I have to do is feed hay and check the cattle." Commercially manufactured bulk feeds gained in popularity during the 1950s and 1960s as farmers relied more heavily on feed prepared with chemical additives.[30]

Augers were one of the most efficient push-button devices on large beef- and dairy-cattle operations. Electric-powered augers could move feed through a tube and discharge the feed from holes in the sides of the tube. An auger system on the Wiemers farm near Melbourne allowed the family to feed 135 head of Herefords in fifteen minutes. The auger system ran from the corn crib and granary, where a homemade mixer (fabricated from an end-gate seeder) combined corn and protein before it conveyed the feed to the silo. At the silo another mixer added silage to the corn-protein mixture before an overhead auger carried the feed to two concrete feed bunks where the feed fell through chutes into the bunk. As each chute was filled, the auger carried the feed mixture farther down the line until all were filled. According to Orval Wiemers, farming eighty acres and feeding livestock with electric-powered systems was preferable to the kind of farming he had done during World War II. He stated, "I was working myself to death," trying to farm between six hundred and eight hundred acres at maximum production. With the new system Wiemers could handle the morning feeding and then take care of his implement business. His fourteen-year-old son did the afternoon feeding after school. The Wiemers family used family labor to maintain a medium-sized herd without hiring workers.[31]

The silo unloader was another tool farmers used to cut their labor demands. Instead of pitching tons of silage down a chute from the top of the silo every year, it was possible to install an electric-powered auger that moved across the top of the forage in the silo like the hand of a clock, shaving off a layer of silage and conveying it to a blower that discharged the silage down the chute. To feed silage to 100 steers by hand a farmer could expect to spend approximately one and a half hours per day in the silo, pitching silage down a chute from the top of the stack. With an auger unloader that job took only thirty minutes. If a family invested twelve hundred dollars in an unloader to feed a herd of 35 cattle, the cost of the machine averaged one dollar a day, which was much less than the cost of wages.[32]

The potential labor saving of silo unloaders and auger conveyor systems was a powerful incentive for farmers who raised or aspired to raise large herds. Francis Kenkel of Shelby County wanted to expand his beef-feeding operation but was concerned about high labor costs. Instead of hiring labor, he installed an automated system that let him feed from 400 to 500 head of cattle per year on his farm. He testified that he could feed 275 animals by himself in twenty minutes. Ovey Vaala of rural Melvin in northwest Iowa converted to an automated system in 1947 in response to his frustration with hand feeding from a wagon. He spent up to three hours per day feeding a herd of approximately 80 head. Furthermore, he incurred the expense of using a tractor for that entire time. Vaala installed an auger-and-tube system over a concrete bunk, which allowed him to feed as many as 170 head of cattle at once. Vaala boasted that he was able to feed twice as many animals in twelve minutes as he used to feed in three hours. In 1959 the value of saved labor alone amounted to approximately two thousand dollars.[33]

The importance of automated materials handling was more apparent as the labor situation worsened in the late 1950s and early 1960s. Jim Cochran of Dallas County complained that "Good, dependable hired help is getting harder to find. So," he concluded, "the best way for me to sidestep this labor bottleneck was to mechanize my feeding." Carl Feucht of Lyon County conceded that feedlot mechanization did not always provide a maximum return on investment, but it was preferable to depending on the vagaries of the labor market. As Feucht put it, mechanization "sure is more dependable than the hired labor situation these days." Norman Erickson of Hamilton County used an auger system to move feed from a mill in his corn crib to an automated feed operation. "Five years ago," he explained in 1960, "we installed the 5-inch auger to move grain and feed fifty-three feet to the feeding setup." When he had to haul feed on a temporary basis during the winter, after years of using the automated system, he realized "how much work the auger had been saving us." Erickson's farm had a capacity of between 200 and 300 cattle, and he kept all of them on the drylot. He claimed that his elaborate electrified feeding system saved him the expenses of hiring an extra laborer.[34]

Farmers even designed their own feeding systems. Joel Middlen of Lyon County endured years of back trouble and decided "I had to start using my head instead of my back." Middlen combined a gasoline engine with parts from a manure spreader and a junked car to create a feed mixer-carrier that permitted him to feed 80 head of beef cattle. Middlen dumped grain and silage into the spreader. The manure spreader's beaters mixed the feed as the moving bed of the spreader pushed the feed out of the machine. An old car top over the beaters kept the feed from being thrown out of the cart, forcing it to fall into the feed bunks below. The mixer-carrier moved out over the feed bunks on a set of parallel tubes, or rails, just as a train moved along tracks. The rims of automobile wheels (without tires) kept the vehicle on the rails. Middlen built a similar mixer-carrier for his brother-in-law. One Hardin County farmer devised a ninety-foot retractable bunk that moved on rails. Homegrown innovation allowed

farmers to use their own time during the slower winter season to minimize capital investments.[35]

In some cases farmers redesigned the layout of their farmsteads completely. Existing buildings formed a part of the reality farmers had to live
with, even when those buildings were not designed with automated feeding in mind. In 1965 Russell Frandson of Story County reworked his farmstead (initially laid out in 1939). He connected the corncrib and a cluster
of three small silos together with overhead augers, which allowed him to
draw on any one of the silos for mixing silage with grain. He tied another
silo in to the system with a second overhead auger. The lines met to carry
feed into two long feeder augers running perpendicular to the main line,
which deposited the feed into two bunks, one 144 feet long and another
105 feet long, where he fed up to 500 head of cattle.[36]

There were significant financial costs to these new systems, yet for
many farmers who had already invested in silos and feedlots in the years
before automation, the cost of mechanization was not daunting. According to one 1959 advertisement, augers and motors cost approximately
$700.00 for farmers who fed fewer than 300 head of cattle. Three years
later Robert Peet of Jones County reported that he had invested $700.00
for an auger system. The augers were the least expensive part of the operation. Costs ranged from $1.50 per foot for 4-inch-diameter augers to $2.00
per foot for 6-inch-diameter augers. Jim Cochran's 160-foot bunk cost
about $5.00 per foot, or a total of $800.00. This figure included the bunk,
the concrete surface for the cattle to stand on while they ate, and the
crushed rock path to keep the tractor and self-unloading wagon out of the
mud. In 1959 Marvin Fernow of Linn County invested approximately
$1,750.00 in electric motors, auger, and tubing to feed his 435 cattle in
less than twelve minutes. Silo unloaders cost from $1,000 to $1,500 in the
late 1950s. These expenses would have been unthinkable to an earlier generation, but for farmers confronted with high labor costs, the expense was
reasonable.[37]

Costs were appreciably higher when farmers created entirely new systems. When farmers created an operation from the ground up, rather than
simply converting an existing feedlot, they confronted the expenses of
erecting new silos, feed bunks, sheds, and fenced lots. This was the kind of
setup that Lawrence and Gordon Hauck of Humboldt County installed
around 1960 to fatten two herds of 350 cattle every year. The most important labor-saving element of their operation was a circular feed bunk with
an overhead swinging auger to carry silage and feed from the corn crib
and three new concrete silos that were eighteen feet wide and fifty feet
tall. In total their system cost $15,000, which included concrete feed
bunks and concrete floor to keep cattle out of the mud.[38]

Farmers with a high tolerance for risk were not intimidated by the cost.
In 1962 Bob Anson of Marshall County invested $25,000 in an automated

beef-feeding operation for 750 head of cattle and 700 feeder pigs per year. His new farm layout included two silos measuring twenty feet wide by sixty feet high, with one-hundred-foot auger bunks. Anson managed all his animals and farmed seven hundred acres with the help of only one hired man. In spite of low cattle prices in the mid-1960s, Anson justified the expense of his operation by comparing the first cost of the feedlot and buildings to the cost of hired labor. "If I didn't have my beef feeding equipment, I'd need another man," he explained. "In ten years, his wages would match my feedlot investment which should give me at least 20 years use." The calculations done this way show that Anson stretched his capital improvements over twenty years, instead of spending the same amount of money on labor over ten years.[39]

Even risk-tolerant farmers were careful about planning, however. The Berghoefer Livestock and Grain Company, located in Franklin County, was a large family enterprise that expanded incrementally in the 1960s. The Berghoefers counseled careful planning based on expert advice combined with observation of existing facilities and other operations. They expanded from 400-head capacity to 800-head capacity in three years. Based on their experiences, they counseled others to do the necessary planning and research and to "put it up a piece at a time to make sure each thing pans out as you go along—missteps are easier to correct early."[40]

Experts encouraged farm families to put pencil to paper and to determine beforehand if automated feeding could be profitable on their farms. The most variable cost was labor. If farmers paid low wages, then it was potentially more profitable to continue hand feeding. If wages were high, then push-button techniques were more attractive. Similarly, if a farmer spent a lot of time feeding animals, labor costs added up more quickly than if a farmer spent comparatively little time in feeding. *Wallaces Farmer* published the results of a Purdue University study on the costs of labor and the profitability of automated feeding. The study indicated that, if a farmer could save an hour a day over the course of two feedings and paid a hired man $0.75 per hour, that farmer could afford to invest $1,750 in mechanized feeding. But a farmer who paid $2.00 per hour could afford to invest $4,800 in order to save an hour a day. These calculations indicated just how rapidly labor costs could add up over the course of a year and proved that capital investment could help farmers. Automated-feeding veterans claimed that their system was more reliable than hired men. In the years since an O'Brien County family installed their system, they had missed only one feeding. "Not many hired men can make a claim like that," they asserted. "Making do" or just getting by with hand feeding was not an attractive option for farmers who hoped to continue farming in the postwar period.[41]

Automated feeding helped farmers deal with the cost-price squeeze not only by cutting labor costs but also by expanding production. The pressure on farmers made standing still in terms of farm practices a risky position. Dairy farmers used automated materials handling and new milking parlors to expand production in order to offset shrinking returns. Similarly, beef producers could automate in order to expand. A cattle feeder from Clinton County, Ferd Schmidt, made an educated guess about the impact of push-button techniques on feeding trends. "If it weren't for mechanization in the feedlot," he asserted in 1961, "I bet there'd be one-third less cattle on feed" in Iowa. "If we had to go back to the scoop and basket, I know we'd be feeding fewer cattle than we are." Schmidt's analysis involved some guesswork, but it reflected an important truth about the 1950s. When it took as much time to feed fifty cattle by hand is it did to feed three hundred by machine, the investment in automation was something that could pay for itself in a matter of years.[42]

Automated feeding also made it easier for people with less skill or physical strength to handle routine feeding. Teenaged, elderly, and disabled workers were less expensive than those in their prime who often had families to support. Carl H. Goeken's system on his Cass County farm was designed to help him do in thirty minutes what it previously took two men half a day to do. "I wanted to have a setup that could be operated by a single hired man or a 'retired' person," which suggests he could not only cut labor costs but also have someone less physically fit do the work. Hired men testified to the decline in physical labor and the relative de-skilling of farmwork. Goeken's hired man claimed that the system was not complicated. "Main thing," he explained, "is just to push the buttons in the right order so that the wrong thing doesn't happen." Pushing buttons in sequence required little brawn or brainpower. Ted Pellett of Cass County claimed that the ease of push-button farming was a welcome change from the scoop shovel. According to Pellett, "It's nice to stand inside and feed cattle when it's pouring down rain or a blizzard." Now farmers who fattened livestock could enjoy the creature comforts that dairy families and workers appreciated when they converted to modern milking parlors.[43]

Through the 1960s, mechanized feeding remained a practice of only a small minority of farmers, however. In 1964, 49 percent of farmers in Iowa reported feeding cattle, while only half of those reported feeding cattle year-round. Those who could justify the expense of automated feeding amounted to only 25 percent of Iowa farmers. Of the farmers who fed cattle, 9 percent used fence-line feed bunks and slightly less than 7 percent used power conveyors such as augers for feeding. The most popular form of labor-saving device in cattle feeding was the self-unloading wagon. One-fourth of all cattle feeders used these wagons to fill feed bunks either along fence lines or, more commonly, in the middle of feedlots.[44]

Even though it was a minority of farmers who fully automated their feedlots, this minority played an increasingly important part in Iowa agriculture. Over the course of the 1960s, the number of feedlots declined, but the share of statewide cattle production of the largest of those feedlots increased. Iowa led the nation in the total number of feedlots for cattle as well as in numbers of cattle marketed. In 1969 there were 44,002 feedlots of all sizes in the state, down from 46,000 in 1968. The decline was in the number of feedlots with less than a thousand head of cattle, although the remaining feedlot operators with less than a thousand head actually marketed more cattle in 1969 than they did in 1968. The number of lots with more than a thousand animals did not increase from 1968 to 1969, although their share of total production for those lots increased slightly. In 1968 the largest feedlots marketed 8 percent of the total number of cattle marketed in the state. In 1969 the largest feedlots marketed 9 percent of Iowa cattle. The shift of 1 percent from one year to the next did not constitute a trend by itself, but the shrinking number of small-scale producers was a major shift in production.[45]

It is almost impossible to know anything about the motivation of the two thousand smaller-scale operators who quit raising cattle from 1968 to 1969, but the example of dairying is instructive. Dairy farmers in large numbers shifted operations in the 1950s, preferring not to invest in the new equipment that made expansion possible. Some continued for a while, but they could not last long producing for a lower-priced Grade B market. There were many thousands of small-scale operators who continued to raise beef profitably, but there were also many who found that the capital requirements of livestock production were too steep to continue, just as dairy farmers had realized. Those who stayed in farming concentrated on other tasks such as raising cash grain. Many who quit farming reached a point in their life cycle where they could not justify new investments, and many others sought opportunities elsewhere.

Confinement

Confinement feeding was a fulfillment of farmers' push-button dreams and the ultimate expression of industrialization in livestock production. In confinement operations, many of the steps in the process of helping animals reach market weight could be mechanized, from feeding to manure removal. This was a significant change from the ways farmers had raised hogs up through the mid-twentieth century. Settlers during the nineteenth century frequently let their hogs roam for much of the year, fencing fields to keep animals out rather than constructing fences to keep them in. By the twentieth century, animals were frequently enclosed on Iowa farms. As population density increased and a handful of farmers began investing in improved breeds of livestock, communities passed new fence laws to keep animals in so as to help control breeding. Throughout most of the twentieth century, farmers pastured hogs in the warm months, turned them to cornfields with cattle in the fall, and moved them onto lots adjacent to the corncrib for fattening before marketing. This enabled farmers to convert corn, a low-priced commodity, into higher-priced pork and lard. Hogs did not require much work when they were on pasture or "hogging down" a cornfield. Farmers hauled drinking water, but the animals typically thrived on the pastures and in the cornfields. Hogs enclosed in lots required much more work, including hauling feed and water.[46]

In the early 1950s confinement feeding simply meant restricting hogs to a concrete floor. Hogs confined to concrete lots conserved body mass, since they were not constantly expending calories to pull themselves out of mud. This was not, however, the most popular method of hog production. In 1955 only 29 percent of Iowa farmers raised hogs on concrete. Most farmers did not believe it was practical to confine animals. As one Ida County farmer stated, "It's healthier to have 'em out on pasture." There was too much risk of disease among hogs kept on small cement-floored lots compared with hogs kept on pasture. The lots provided a disease-rich environment, with hogs in close quarters much more susceptible to communicable diseases than those on pastures.[47]

Three developments made confinement feeding more practical in the mid-1950s. The first was that antibiotics became available to prevent or control disease outbreaks. The second was the increase in farmland prices. Land prices increased in part due to the competition of expansion-minded farmers who wanted to gain economies of scale with new labor-saving equipment. The third development was that stagnant prices for farm commodities meant farmers needed to obtain maximum cash returns from each acre. Pasture-grazing hogs on large tracts simply did not pay returns as high as corn or soybeans. Albert Kolder of Butler County stated bluntly, "My fields are worth more to me in crops than in pasture." Dairy and beef farmers learned that it was more profitable to keep animals on lots and haul feed to them than to let the animals graze, because the rate of gain was faster with controlled feeding. As land values appreciated, farmers wanted to get the most possible production from their land, which often meant raising more grain and forage or a high-return crop such as soybeans. Confining hogs to cement feeding areas allowed farmers to get the most from their land. It also allowed farmers to market hogs throughout the year instead of selling only in January and February when prices were at their lowest point of the year because everyone else was also selling hogs at that time.[48]

The disadvantage of confinement feeding was the high labor requirement. Instead of letting hogs graze for themselves and deposit their manure across a large acreage, farmers who confined hogs to concrete lots needed to haul feed to hungry hogs and move tons of manure, both of which were labor-intensive jobs. Oscar Torneten of Pottawattamie County emphasized that it not only took labor to move the manure but it also took time to clean and disinfect the feeding equipment used in the lots to prevent the spread of disease. It was not possible to continue to use the scoop shovel and manure fork with confinement feeding.[49]

A small number of farmers experimented with automation and architecture to reduce the labor demands of confinement feeding and eventually changed the definition of confinement. Instead of simply confining hogs to lots with hard-surfaced floors, confinement came to mean confining hogs to pens within a building. These pole-framed buildings were divided into pens of varying size. Max Bailey of Story County constructed a new building along these lines in 1957. A feed mill and storage bin occupied one end of the rectangular building. An auger system carried feed down the center of the 176-foot-long building and deposited feed along one side of the eighteen pens. These 16-by-18-foot pens included an automatic watering device for fresh drinking water. Lengthwise along the exterior walls there were dunging alleys. During normal operations, a portion of the alleys were included in each pen, but movable gates could close off each pen and open the alley from one end of the building to the other. A tractor loader could scrape the manure off the floor and push it into a pit at one end of the building. The new style of confinement feeding was a completely integrated system designed to handle large numbers of animals with minimal labor.[50]

Manufacturers offered prefabricated confinement-feeding structures such as the Bacon Bin, manufactured by a company in Kansas City, Missouri, that resembled a steel grain-storage bin with corrugated siding and a conical roof. The interior included two levels of pens arranged around the circumference of the structure, which could be rigged for farrowing or finishing hogs. In addition to featuring automated feeding, watering, and waste removal, the building was completely insulated and included a ventilation system and temperature control. William Conover of Marshall County installed a Bacon Bin on his farm in 1964, one year before the structures came on the market. He claimed it was an inexpensive structure to build and that his hogs gained weight rapidly with minimal labor costs.[51]

Iowa hog farmers readily substituted capital for labor in livestock operations. Mearl Pottroff of Carroll County constructed a 36-by-100-foot building with twelve pens to accommodate up to fifty hogs per pen. Like Max Bailey, with this kind of operation, he could feed many more hogs than was possible with older systems. By investing approximately $4,500 in materials and an undisclosed amount for labor, Pottroff attempted to produce a large number of market hogs as rapidly as possible. USDA economists estimated total costs for a five-hundred-hog operation to be approximately $10,000. Other farmers such as Don Goodenow of Ida County spent considerably less to convert an old barn to a confinement barn. He installed separate augers to move feed into the barn and manure out of it. He claimed that he spent approximately one hour a day to chore and inspect four hundred hogs. E. L. Quaife, an Iowa State swine specialist, believed that these new facilities were best suited for farmers who were willing to raise more than five hundred hogs per year. With large investments and large-scale production, confinement feeding was the choice of a few rather than of the many.[52]

Robert Hamilton of Hardin County was another of those rare farmers who moved his animals off pastures and into buildings and redefined hog production. Hamilton confined not only the hogs he planned to sell but also his breeding herd. In 1957 he constructed a $20,500 farrowing house and in 1959 invested $4,000 in a confinement building to finish hogs. But the automatic feeders, water system, and manure gutters were not the unique features. Hamilton's system was unique because he kept hogs inside all year long. Hamilton surmised that the labor savings realized by farmers who housed animals inside for part of the year could be realized the entire year.[53]

Beef producers also wanted to realize the improved rate of gain that hog producers enjoyed from confinement feeding. Some beef producers already automated their feedlots, but few farmers provided permanent year-round shelter for their beef cattle. Cattle needed shelter only in the most extreme weather, but providing any kind of shelter helped them convert more of their feed to weight gain instead of just using it to keep warm. For most farmers during much of the twentieth century, however, potential gains were not worth the expense. Simple pole sheds were the rule for farmers who wanted to provide protection for their beef cattle. A Taylor County farmer did not have any structures designated as cattle shelter. He explained, "My cattle spend the winter in the grove and around the buildings" to escape the wind and cold. In the 1960s a small group of cattlemen and farmers hoped they could get more gain by controlling the environment for livestock, just as hog farmers had begun to do in the 1950s.[54]

Farmers who confined beef animals could minimize the extremes of temperature while gaining all the benefits of automated feeding, watering, and manure removal. The most publicized cattle confinement in the state belonged to Jim Rock of Kossuth County. Rock constructed his 72-by-160-foot building in 1961 at a cost of $1.25 per square foot, including labor charges. For three hundred cattle he used cornstalks and corn cobs for bedding over a concrete floor. The bedding kept the cattle dry and out of the manure, while the packed manure generated heat in the winter time. A fence-line bunk along one side of the building allowed for automated feeding. In 1962 Rock reported that he was pleased with his operation, but he would not know how successful it really was until he was able to use it for at least another year.[55]

After several years of operation Jim Rock and other farmers who had chosen confinement praised the system. Rock moved cattle weighing from 850 to 900 pounds into his building and kept them there until they were ready for market weighing between 1,050 and 1,150 pounds. He concluded that the barn "really pays off in this final push." The cattle in his outside lots gained about 0.10 pounds a day less than his cattle in confinement. This small margin, spread over three hundred cattle every year for several years added up quickly. After nine years of operation he calculated the barn had paid for itself in five years. Robert and Roger Clause of Grand Junction agreed that confinement feeding paid. They built their facility in 1968 when they hosted the Farm Progress Show, an annual farm exposition sponsored by *Wallaces Farmer*. While they would not divulge the specifics on the profitability of their operation, they expected that the building would pay for itself in ten years or less. Some of the most persuasive evidence that beef confinement could work was the fact that Rock's brother-in-law copied this system in 1969, and in 1970 Rock planned to build another barn like his old one.[56]

There was no consensus among farmers that confinement was the best solution for finishing cattle, even as a handful of farmers experienced success. In 1970 a new partial-confinement operation opened near Aurelia in Cherokee County to accommodate twelve thousand steers but this was exceptional. Farmers who voted with their money voted against the new system. There were fewer confinement operations for cattle than there were for hogs. H. L. Self, an animal scientist at Iowa State, argued that, to justify the capital expense, confinement feeders needed to earn at least sixty dollars more per steer than those who did not use any shelter at all. Self acknowledged that this was a difficult goal to reach and implied that most farmers would be wise to continue traditional feeding practices.[57]

Experts such as Self played an important role in promoting confinement feeding for hogs and cattle. Livestock specialists at Iowa State and the extension staff studied the proper size of pens, construction materials, ventilation, and arrangement for manure-disposal systems. Some of these experts were on hand at the 1962 Farm Progress Show where Iowa State dedicated its exhibit to hog production and called it "Pigneyland." Pigneyland focused on creating the ideal type of hog carcass for modern consumers and the architecture and equipment needed to raise hogs from farrowing to finishing. In the finishing building, the emphasis was on pens with slotted floors for keeping pens free of manure. Extension specialists answered questions from interested farmers. Crowds were always large for these expositions, but the large crowds and the interest expressed in learning more about the new technology do not indicate that a majority of farmers were willing to commit to confinement.[58]

The reality was that few farmers raised hogs in confinement during the 1960s. In a 1966 poll of Iowa hog farmers, only 2 percent of respondents stated that they raised hogs in buildings with insulation, ventilation, and automated feeding or manure-handling systems. Most of the interest in getting into modern confinement feeding was shown by farmers under the age of fifty, and 9 percent of this group was interested in confinement feeding while only 2 percent of the over-fifty group indicated any interest in it. There were many farmers who profitably raised several hundred head of hogs on pasture during the 1960s. Roger Shaff of Clinton County converted to confinement feeding but insisted that it was not for everyone. Farm families with capital and little labor could make it pay, "But if you have enough labor and not a lot of capital, there's no advantage." The handful of farmers who chose to use confinement feeding during the 1960s suggests that farmers regarded it as unproven, unwise, or unnecessary.[59]

Farmers who continued to raise hogs on pasture explained their decisions to stay with their system. They reported good results with pasture systems, contending that they kept animals on clean ground and did not have to haul manure. Some farmers were content with the way they did business. A farmer from Plymouth County asserted that he would never abandon pasture feeding. "I raise over 500 head of hogs in the field and it's simple and easy." For others, prior investment in pasture feeding meant that there was little incentive to change. Older farmers lacked the flexibility to invest due to anticipated retirement. "I'm not as young as I used to be," explained Don Kruse of Floyd County, "and I can't afford to invest that kind of money at this stage of the game." Renters avoided investing thousands of dollars in permanently situated confinements, although there were a few exceptions. As one renter commented, "It would make a difference too, if you were just in the hog business. I feed about 140 head of cattle a year too, and I have a lot of money tied up in them." This commitment to diversification was an important anchor for farmers who hoped to ride out low prices for one commodity with receipts from another commodity.[60]

The high first cost of constructing a confinement system could be profitable, but that kind of investment necessitated expanding production. A farmer from Washington County advised a farmer who wanted to raise a herd of five hundred hogs that a confinement building "would not be profitable unless you increase your hog production." While only one Iowa farmer in twenty raised more than five hundred hogs per year, the largest group of producers, 28 percent of farmers, raised between one hundred and two hundred head. Furthermore, the percentage of sales of hogs for producers in this category increased over the course of the early 1960s. The growth in this group suggests that most farmers liked raising hogs on pasture with family labor and that they were more risk averse than the handful of farmers who invested in confinements.[61]

Farmers who utilized confinement feeding prized the potential for expansion. These risk-tolerant farmers wanted to invest in systems that would enable them to ride out future price shocks. In 1965 a Boone County farmer observed that confinement was gaining in popularity. "Oh, it's coming, all right," he noted. "There aren't too many in our area yet—but it's coming." In the mid-1960s Carl Frederick of Johnson County was raising sixteen hundred hogs per year. He maintained that farmers needed to fatten hogs year-round in order to make their investments in buildings and equipment pay off, and confinement was absolutely necessary. Frederick explained, "I feel that hogs will be in controlled buildings—they'll never leave buildings from birth to market." By controlling feeding and by minimizing variables such as fluctuations in the weather, industrial-style production allowed farmers to be more profitable than producers who used pasture systems.[62]

Industrial-style production was profitable but not necessarily problem-free. The idea behind confinement feeding was to speed the weight gain in animals by removing variables that caused stress. In reality, confinement caused its own stresses, which resulted in undesirable animal behavior or illness. Veterinarians reported that farmers who used confinements observed more problems with aggressive behavior such as tail biting. The concrete floors took their toll on animal health, causing sore feet and legs. Some veterinarians argued that mineral deficiencies in confinement-raised hogs contributed to tail biting and lameness and that farmers needed to increase levels of salts and trace minerals in the hogs' diet. If ventilation was not adequate, the dust of feed and waste as well as airborne pathogens could cause respiratory ailments. One Oelwein veterinarian summed up the situation by observing that "practically the same disease problems [exist] in confinement as in pasture systems, except that I find more of them and more that are difficult to treat."[63]

Veteran confinement feeders and experts counseled that improved herd management would remedy these behavior and medical problems. Hogs on the pasture system took care of themselves as long as farmers provided feed and water. Confinement feeding was different and required that highly skilled labor substituted for merely skilled labor. The old push-button image of farmers relaxing while machines did the work was illusory. A beef-confinement feeder maintained, "There's a lot more to it than just pushing buttons." Farmers always had to watch their herds, but with the increased investment in buildings and automated systems and with larger herds, they now had to pay even more attention to the details. In some operations both workers and visitors showered before entering livestock areas. Washing machines and dryers for workers' clothing were new and necessary tools to prevent introduced pathogens from infecting the herd or from leaving the property. It was critical to monitor nutrition, equipment, ventilation, health, and waste management. Farmers who did not perform this high degree of oversight risked seeing one small problem magnified many times into a major problem.[64]

Manure

Waste management was one management detail that assumed monumental importance for confinement feeders. Large-scale confinement feeding brought a large-scale manure problem. Hogs on pasture deposited urine and feces across many acres. By contrast, hogs in confinement buildings deposited their waste in a small area. The amount of waste excreted by one hog on full feed in confinement was staggering. A 100-pound hog that consumes approximately 4.7 pounds of feed per day produces 4.3 pounds of solid waste and 1.06 gallons of liquid waste per day. As the hog reaches market weight of 250 pounds, it consumes 8.2 pounds of feed per day and excretes 7.8 pounds of solid waste and 2.65 gallons of liquid waste per day. In a confinement barn, 250 hogs generate almost 1 ton of solid waste and approximately 662 gallons of liquid waste per day during the period immediately prior to sale.[65]

Confinement feeders needed new ways to deal with these massive amounts of manure. They constructed manure pits under the confinement buildings to collect solid and liquid waste. Large-scale operations that included automated feed, water, and ventilation systems also included automated manure handling. Herman Tripp of Greene County constructed an underground manure pit made of concrete blocks that measured five by six by eight feet. Each week he used a four-inch auger to transfer the manure from the pit to a spreader wagon that carried the waste to his fields. These pits required regular labor to clean. If they could not be emptied for some reason, the farmer faced a manure backup of industrial proportions.[66]

The manure lagoon was a labor-saving solution. A lagoon was simply a large earthen basin or pond to hold animal waste and to allow the waste to decompose before removal and use as fertilizer. There were two types of lagoons, aerobic and anaerobic. Aerobic lagoons held enough water to support bacteria that required oxygen; anaerobic lagoons were so concentrated that only bacteria that lived in environments without oxygen could thrive. Research from the mid-1960s indicated that most lagoons had both aerobic and anaerobic qualities, although experts and farmers believed that the wide shallow lagoons were aerobic while the deep lagoons were anaerobic.[67]

In the early 1960s farm families who practiced confinement feeding experimented with lagoons. Robert Schrier installed a sloping concrete gutter to channel manure to a lagoon on his Cass County farm. Every day he scraped the hog waste into the gutter and every week or ten days he removed the plug on the gutter and gravity would carry the waste through an eight-inch tile line into his lagoon. This technique allowed him to remove manure from a 38-by-80-foot building that housed between 250 and 300 hogs. Farmers with manure collection pits had to drain those pits on a regular basis. Sometimes this was every week, other times every few months. By contrast farmers with lagoons were free from the regular and frequent round of moving manure.[68]

Experts played a larger role in farming operations as farm families attempted to deal with waste. With the large sums of money involved, sometimes tens of thousands of dollars for livestock barns, grain-handling buildings, and waste-management facilities, consultants helped farmers maximize returns on their investment. The extension service and Midwest Plan Service provided free or low-cost advice and plans for farmers who constructed waste lagoons or new buildings. The Midwest Plan Service, headquartered in Ames, was a consortium of agricultural engineers who drafted farm-building plans. The Cooperative Extension Service conducted research on the effects of developments such as paved lots and shelter on the rate of gain in feedlot animals. Publications of both the extension service and the Midwest Plan Service showed the growing complexity of farm planning, especially for livestock manure.[69]

By the end of the 1960s farmers who used new manure-handling technology were confronting unanticipated problems. The manure lagoons and manure tanks posed risks to both farmers and livestock. There had always been discussions concerning the smell of manure pits and lagoons, but the toxic gas they released became a real threat to the safety of farm families. Stored liquid manure produced both methane and carbon dioxide, which were released as farmers pumped out the manure pits. Without adequate ventilation these gasses could asphyxiate humans and animals. Ammonia, a respiratory tract irritant, and hydrogen sulphide, which causes severe headaches, dizziness, and even death, built up to dangerous levels in poorly ventilated structures. Lagoons also proved complicated. Farmers could not simply create them and walk away. Experts advised farmers to create a series of lagoons, including an anaerobic lagoon for solid wastes to settle, followed by an aerobic lagoon. In some cases experts recommended that farmers line the lagoons with clay to reduce groundwater contamination. The thick sludge had to be removed in order to keep the bacteria in the lagoon working to break down the waste. Sludge buildup prevented decomposition, which resulted in groundwater contamination or runoff into surface waterways.[70]

Agricultural engineers from Iowa State corresponded with farmers and conducted occasional site visits to help them implement best practices for safety and manure issues. Extension agricultural engineer Dale Hull corresponded with several farmers and farm managers over the course of the 1960s and counseled them on techniques to minimize the manure runoff into watercourses. From June through September of 1966, Hull and his fellow extension agricultural engineer Ted Willrich advised Rod Lorenzen, the manager of a livestock production company called Group 21 Incorporated. Lorenzen initiated contact with Hull for assistance in preventing the contamination of Waterman Creek in O'Brien County. Hull recommended the construction of a collecting basin in which runoff and solid waste could settle. Any overflow drained into a shallow two-to-three-acre oxidation pond. Three months after making these recommendations, Hull and Willrich flew to Moneta "to take a good look at your manure disposal problems in connection with pollution control on Waterman Creek." Shortly after making the trip Willrich submitted an evaluation of the feedlot site and provided two contacts for engineering services.[71]

Willrich advised the feedlot manager that waste runoff from thirty acres of lots and roads entered Waterman Creek. Willrich warned that "The Iowa Water Pollution Control Commission would no doubt look unfavorably on the continued operation of this feedlot until such time as a satisfactory runoff treatment system or runoff utilization system was installed." Willrich's recommendations were more complicated than those suggested by Hull in June. If Group 21 managers wanted to utilize their manure as fertilizer, they could prepare a system of settling basins and install a sprinkler system to carry and deposit wastes onto cropland. The other option was to create a treatment-and-disposal system, which would include a settling basin, an anaerobic lagoon, an aerobic lagoon, and finally an evaporation cell for removing the last of the liquid from the animal waste. Furthermore Willrich advised study of the potential site for the abatement system. He noted that the site should be mapped, tests conducted for soil type and permeability, and flood elevations noted to prevent pollution. It is unclear if Lorenzen had the power to implement these recommendations, but they represent the most current thinking of the experts about livestock-waste abatement at that time.[72]

Farmers were not of one mind about large-scale livestock production, waste runoff from feedlots, or the role of government regulation in stopping pollution. One-third of all farmers polled in 1968 about how to curb runoff from feedlots stated they were not sure what the proper solution should be; 29 percent of the farmers surveyed believed that only the large operations were responsible for any pollution problems and that the majority of farmers were not to blame and therefore should not be regulated. Only 10 percent of farmers believed agricultural-pollution-control measures were not needed. Farmers agreed that feedlots posed environmental problems; 90 percent of farmers believed something should be done to control runoff pollution from feedlots, but they did not agree on who was most responsible for that pollution and who should be regulated.[73]

The differing viewpoints reflected divisions within rural society, differentiated by age, education level, and income. Almost half the farmers under the age of thirty-five wanted more pollution control, while only 33 percent of farmers aged thirty-five to forty-nine wanted more control, and only 19 percent of those from fifty to sixty-four agreed that control measures were necessary. Approximately half of those who had attended college favored regulations on all livestock producers while only 21 percent of those with eight years or less of formal education agreed. Producers with the highest income favored regulation for all producers, while those with gross income of between ten thousand and twenty thousand dollars believed that only the largest producers should be held accountable for pollution control. One respondent made a vague claim that nature solved pollution problems, while another argued that feedlot waste was not responsible for water pollution, because "Water is purified after it soaks thru a few feet of soil." Comments by farmers who favored regulation suggest that their principal motivation was a "not in my backyard" concern rather than the issue of water quality for the people of Iowa. One Greene County farmer exclaimed that "Something has to be done. These large feedlots are a stinking mess." Another Greene County man complained about hog confinements, claiming that "these hogs raised in confinements are creating a problem for folks that have to live next door." One farmer who raised hogs in confinement noted that his neighbors and friends made "kidding" comments about the smell of his farm. He believed that at some point they might do more than joke with him. The threat of regulation, court action, or retribution over the issue of environmental quality was becoming a reality of agriculture in the late 1960s.[74]

Farmers and the general public were more sensitive to the thought of harming ecosystems than had been the case in the 1950s. In the spring of 1968 the Iowa Water Pollution Control Commission called a series of public hearings in Iowa City, Ames, Atlantic, and Storm Lake in order to gather public input concerning the establishment of regulations for the disposal of cattle-feedlot waste. The commission defined a feedlot as an enclosure or group of enclosures in which animal density was greater than fifty head of cattle per acre. Proposed regulations included a permit for any operation of more than a thousand head of cattle, any feedlots near watercourses, and those where runoff or overflow from a manure lagoon or tank might flow onto another's property. While there were very few operations of more than a thousand animals, the other criteria applied to many farmers with feedlots. The Iowa Livestock Feeders Association, the Iowa Beef Producers Association, and the Iowa Farm Bureau Federation opposed the proposed regulations. Durward Mommsen, chairman of the Clinton County Farm Bureau livestock committee, urged a delay of at least two years to study the issue before enacting new regulations. Peter Vos, the Mahaska County Farm Bureau livestock committee chairman, reminded the public and the Water Pollution Control Commission that farmers earned small profits: "Any more expenses and we're operating at a loss." A USDA official advised farmers at the Iowa Pork Industry Conference that they could face legal actions. He cited cases from Texas and New York in which farmers' courts fined farmers tens of thousands of dollars in fees and damages from nuisance suits resulting from odor, noise, dust, and insects.[75]

In 1969 the Iowa legislature amended its water pollution control law to require the registration of large-scale livestock producers. The legislature defined large-scale producers as feeders with over a thousand cattle and a population density of one animal for each 600 square feet of lot. As proposed in 1968, the 1969 law also required registration for feedlots if there was a 3,200-acre watershed above the feedlot; if feedlot runoff or manure overflow entered an underground tile line, well, or sinkhole; or if the distance of the feedlot from a stream was less than two feet for every animal on the lot (that is, five hundred head of cattle had to be at least 1,000 feet from the stream). How many farms met these conditions and were required to register is unknown, but estimates indicate that in 1969 only 135 of the 46,000 beef feedlots in the state claimed over a thousand cattle.[76]

State regulators alerted feedlot operators to register in the fall of 1969. By July 1970, 52 cattle-feeding operations had registered with the Water Pollution Control Commission, and 53 more had notified the group they did not need to register since they did not meet the registration criteria. State Health Department inspectors visited 16 feedlots and determined that 11 of those had potential water-pollution problems. One year later a total of 147 cattle feeders and 29 hog operations registered with the commission.[77]

The Water Pollution Control Commission continued to inspect feedlots and to require permits in 1971, even though it appeared that only a minority of those required to register actually did so. For farmers this meant more government involvement in their business. A farmer who wanted to create a feedlot during the 1950s and 1960s simply built one. Now, the process involved agricultural experts and bureaucrats, and construction delayed for months. First, the farmer consulted with an extension livestock specialist at a regional office for advice on planning the layout of the lot or confinement and the waste-control systems. The next step was to contact the Iowa Geological Survey office in Iowa City to determine if the site possessed enough underground water and that the water was of the proper quality to sustain the planned operation. U.S. Soil Conservation Service (SCS) specialists provided advice on facility design. The local Agricultural Stabilization and Conservation Service was the contact for federal matching funds through the Rural Environmental Assistance Program (REAP). After approval by the SCS, the farmer could then register with the Iowa Water Pollution Control Commission. Operations situated in a flood plain or those that required more than five thousand gallons of water per day required further registration with the Natural Resources Council. Registration and approved plans in hand, farmers could begin construction.[78]

Public criticism, new laws governing feedlots, and concerns about future government regulation of water pollution from feedlots and confinements prompted some farmers to examine and critique their operations. In 1970 Roy Olson of Dickinson County explained that he began to consider the problem of runoff control when some downstream neighbors saw some dead fish and speculated that they were killed by waste runoff from Olson's feedlot. Regardless of whether the wastes from his farm contributed to the fish kill, Olson recognized the need to change. Arnold Olson of Emmet County had no abatement plan whatsoever and knew that solid and liquid waste simply washed off his feedlot into the county road ditch, through a culvert and into a nearby lake. Both farmers wanted assistance.[79]

These farmers implemented new waste-management plans. Arnold Olson brought in local SCS workers and Ted Willrich of Iowa State University extension. They designed a collecting basin at the base of the feedlot to hold five inches of runoff from his lot. As solid wastes settled, Olson pumped the liquid out and spread it on his fields. He installed an eight-inch-diameter underground tile line to drain remaining liquids into his field. Olson planned to dredge the solids out of the basin every few years. Roy and Scott Olson fed two thousand cattle on their twelve-acre feedlot that was near the Little Muddy Creek in the Little Sioux River watershed. The site was below fifteen acres of watershed, which drained across the feedlot. Rainfall and snowmelt picked up all the solid and liquid waste and carried it downstream. To stop water from crossing the lot they installed a twenty-four-inch tile line to divert water before it crossed the feedlot. Water falling onto the feedlots themselves was also a problem. They constructed a collecting basin to hold water from the lot as well as small check dams along waterways to catch solids and let them settle out before the runoff continued to the basin.[80]

In 1970 Dale Hull and Stewart Melvin, another extension agricultural engineer, traveled to Peterson, Iowa, to inspect the farms of Don Plagman and Walter Ankerstjerne. Both farmers wanted to construct beef confinements on their respective farms. Melvin suggested that both farmers construct oxidation ditches to carry wastes into a series of two anaerobic lagoons. Melvin also predicted there would be a possible odor problem for a neighbor located approximately one-quarter mile away from the proposed Ankerstjerne farm site. To prevent contamination at the Plagman farm, Melvin suggested he should bank his lagoons and construct them partially above ground due to the high water table.[81]

Farmers who constructed waste-control systems hoped to head off criticism, prosecution, and regulation as well as to be good neighbors. When Bill Conn of Kossuth County planned his 2,000-head feedlot in Kossuth County there was concern in the community that his livestock waste would drain directly into the nearby Des Moines River. Conn noted that a local women's group came to his farm to investigate and to ensure that his lots would drain away from the river toward a retention pond. As Roy Olson explained, "We don't want anyone pointing a finger at us for causing dirty water." Similarly, Arnold Olson stated that "Clean water, clean air, and clean environment is really the thing right now. I would surely advise anyone building a new feedlot to consider runoff controls." Costs for waste-control measures ranged from one dollar to six dollars per head of cattle, depending on local conditions and herd size. Many farmers were willing to pay these costs, especially with assistance through the REAP program, because they wanted to avoid both real and perceived pollution problems.[82]

Farmers committed to highly automated and large-scale operations who were subject to the new environmental regulations were the minority in 1972, but the new techniques were important to a growing number of them. Farmers who used push-button techniques accounted for a growing portion of Iowa's farm output. *Wallaces Farmer* recognized the ways in which farmers altered production strategies by sponsoring the annual Master Farmer Program to honor farmers for excellence in farm management and community service. Each year the editors solicited nominations from people across the state. A panel consisting of representatives of the Master Farmer Club, an editor from the magazine, and an Iowa State farm management specialist selected several farmers among each year's nominees to receive the honor.

The men who received the Master Farmer honor were all large-scale producers who used the latest technology. All five recipients in 1970 practiced some kind of automated materials-handling technique, including bulk tanks, outdoor auger feeding, and confinement feeding for hogs. One farmer piped water into feed boxes in his milking parlor to wet the feed to aid digestion. Another farmer used a bulk-feed-delivery service for his hog-finishing confinement. Not surprisingly, one of the five men even designed his own system, incorporating an old coal stoker from a furnace to add supplements to his feed. The commitment to innovation through labor-saving devices, specialization, and expansion was part of Iowa's remade agricultural landscape. This generation's legacies were larger operations, more intensive livestock production, serious environmental consequences, and government regulation relating to waste management on modern farms.[83]

Making Hay the Modern Way

In the years up to 1945, making hay was physically demanding, time-consuming, and labor-intensive. Most farmers cut and raked their grass or legume forage crop with mowing machines and rakes, picked it up from the field with a hay-loader, and hauled it to the barn where they used a horse hayfork to put it in the barn loose, just as it came from the field. This system required three or four laborers who handled the crop at least three times to get it into the barn—once from the windrow to the hayrack, once from the hayrack to the barn, and once in the barn to spread the new hay across the mow. Some farmers baled hay or straw using a stationary bale press or one of the newer pickup balers that could be pulled through the field, but both of these machines required a tractor driver and two operators on the baler to tie the bales using lengths of wire. The most common use of the baler was to bale straw for use as animal bedding. Baling hay was advantageous only for farmers who wanted to sell or move hay during the winter months, since it was easier to handle and stack it for the second time without losing nutrient-rich leaves. For most farmers, baling hay did not pay.[1]

After World War II, with the introduction of new machinery that automatically baled hay or chopped it into small pieces, Iowa farmers reorganized the hay harvest. They now had to choose how to deal with one of the most important crops in their crop-rotation systems and production. They could continue to use the tools that they, their parents, and grandparents had always used, or they could try something new. While a handful of farmers continued with the old techniques, most of them decided to use the new technology for the hay crop. They did so, however, in different ways. All farmers cared about costs of production, labor scarcity, and the quality of the crop, but the tools they chose reflected different goals. Some farm families used forage choppers and hay balers to help them expand their operations or to specialize, in an effort to beat the postwar cost-price squeeze. Forage choppers eased the physical labor of forage production, although chopping was still time-consuming. Hay balers were useful for other farm families with a diverse mix of crops and livestock. While making bales was easy, handling them was still hard work, but new technology helped farmers with diverse goals meet their needs.

Haymaking changed very quickly after the development of the self-tie baler and forage harvester. In the late 1930s a man from New Holland, Pennsylvania, developed an automatic twine-tie baler, which the New Holland Machine Company produced in large numbers beginning in 1940. In 1944 the International Harvester Company produced its first version of the automatic twine baler. Field forage harvesters, also known as forage choppers, were also new in the 1940s. These machines cut and chopped the hay into small pieces in the field and blew the cut forage into a truck or wagon pulled behind the harvester. Farmers then hauled the chopped crop and fed it to livestock or stored it in a barn, shed, or silo. Engineers developed these machines in the 1930s, but few of them saw actual use in fields during the depression years or during World War II because of the low commodity prices of the 1930s and the production restrictions during the war. When domestic production resumed after the war there were two new options to consider when making decisions about handling hay.[2]

Saving Labor, Improving Quality

In the late 1940s Iowa farmers began a rapid switch to new haying techniques, especially baling. In an effort to do more work more rapidly and with less strenuous effort, farmers began "Making Hay the Modern Way," as one journalist noted in 1946. Farmers who were accustomed to the hay-loader or the two-man hand-tie baler complained that these methods involved too much work, especially when they were aware of the labor savings of the automatic twine-tie baler and the chopper. They considered investing money in more expensive machinery to reduce physical exertion. By the time of the 1951 haying season, 41 percent of Iowa farmers planned to bale all or part of their crop, while only 14 percent of farmers still planned to handle their crop as loose hay the traditional way. The rest of Iowa farmers surveyed were somewhere in the middle, with plans to bale varying proportions of their crop. Only 17 percent of farmers planned to field chop their hay. The next year's survey showed that baling was the most popular way to handle the hay crop, with 55 percent of farmers planning to bale their entire hay crop and the commitment to chopping at approximately the same level as 1951.[3]

Farmers made this rapid turn toward new haymaking techniques as they confronted the looming postwar labor shortage. Iowa farmers were confident that they would soon have more labor-saving machinery available to help them do what they had done with human labor before the war. But even though the new machines were available for purchase or hire after the war, farmers for years continued to struggle with the labor shortage. One farmer from Wright County testified that two of his sons entered the army during the Korean War and that he would have to hire more custom work. He anticipated he would "sell the hay in the field and let somebody else put it up." Farm families for the next twenty-five years continued to deal with fewer hired men, smaller family size, and a contraction in the number of farm operations.[4]

Manufacturers and advertisers appealed to farmers' anxieties about the labor shortage as well as their optimism about technological solutions. A 1958 advertisement for John Deere haying equipment focused on labor savings by invoking concepts of modernity, cost reductions, and freedom. Farmers could "Make Hay THE ONE MAN WAY" by using the "Revolutionary" new Deere haying system. A "one man crew" could use a combination mower and conditioner to get the crop on the ground, rake it, bale it, and load the wagon with an automatic bale ejector, the centerpiece of the advertisement. The bale ejector pitched the newly made bale into the wagon hitched to the rear of the baler, replacing the one or two people who used to stack the bales on a hayrack. The other revolutionary element of the system was the automated storage system. A mechanical conveyor moved the bales from the wagon into the barn and along the ridge of the barn before dumping them into the mow. Automating the work of hay storage replaced one or two workers who stacked the bales in the barn, a hot, dirty, and physically demanding job. This revolution would enable farmers not only to profit but to "find new freedom," an alluring claim in a time of labor shortages.[5]

Baling saved labor, but chopping hay was even less labor-intensive. University of Minnesota researchers determined that farmers spent approximately 2.2 hours to put up one ton of hay using the old hay-loader method, 1.7 hours with a self-tying baler, and 1.2 hours with a field chopper. A married couple from Jones County firmly committed to livestock production proved that a husband-and-wife team could make 80 acres of chopped hay by themselves without hired help. Only 17 acres of their

240-acre farm were planted in corn while 80 acres was in a mix of alfalfa and brome hay that they used to feed beef cattle. They used a mower with a crusher attached to cut and crimp the stems to speed drying. Then the wife operated the chopper while her husband hauled the hay to the barn and unloaded it. By substituting machinery for a crew of three or more workers this couple reduced labor costs. Furthermore they spared themselves a great deal of demanding physical work. They performed all the work from the tractor seat, with automatic unloading wagons.[6]

A rapid hay harvest was important because harvest coincided with the labor-intensive work of cultivating the corn crop, which gave farmers more incentive to bale or chop. While those who used newly developed growth-regulator herbicides such as 2,4-D could expect to reduce the amount of time they spent controlling weeds in their cornfields, only a few farmers abandoned cultivating. During fair weather most farmers continued to face the difficult choice between cultivating or haymaking. Almost half of the farmers surveyed in 1957 agreed that they cut hay when it fit in best with other work. An Ida County farmer stated, "Hay making has to wait on corn cultivating on our farm. Corn is the important crop." If farmers could find a way to speed haymaking, they could lessen the pressure to choose between two or more important tasks.[7]

Hay crushers or hay conditioners were also valuable tools in speeding the harvest. A conditioner consisted of a set of rollers (either smooth or fluted) mounted on a frame that could either be pulled on its own or behind a mowing machine. Each pair of rollers squeezed or crimped the stems and leaves of the crop, which accelerated drying, since the moisture from the thick stems evaporated faster after the crimping process. One Marshall County cattle feeder with ninety acres of hay used his mower with a conditioner in the 1959 season and claimed that crushed hay dried in half the time it took normally in a windrow. "Last year was the first we'd ever tried a hay crusher," reported Earl Felt, a dairy farmer from Dallas County. "It eliminated at least one day's drying time." Hay conditioners became especially popular in northeast Iowa where commercial dairying was more widespread.[8]

Drying was a critical issue for farmers, and it became more complicated as they started to bale hay. Every farmer knew someone or had heard stories about someone who had stored hay too moist. At best moist hay can become moldy and unpalatable to livestock. At worst moist hay can spontaneously combust and burn down a barn. Microbes that break down the carbohydrates in the plant material are killed off by the heat generated by tightly packed moist hay. Once these microbes are dead, a chemical reaction occurs that generates flammable gasses, which ignite when they contact oxygen. As long as farmers stored loose hay, they could put it in the barn with moisture content from 25 to 28 percent. Baled hay, however, is more compact and has less chance to dry. It can be stored safely with a moisture content of no more than 25 percent. Hay conditioners were important implements for farmers who baled, since they allowed a faster harvest and minimized the risks involved in using the new baling technology.[9]

Even though speed and labor savings were important considerations, farmers also needed to consider quality. Baling offered labor savings, but farmers who chopped hay or made haylage (a fermented product like corn silage) reportedly made better-quality feed. By the 1950s there were reports that chopped hay and silage retained more vitamin A and protein than baled hay. Both nutrients were critical to livestock growth. Iowa State and experiment station tests indicated that as much as 15 percent more protein and 10–20 percent more vitamin A could be retained in the plant material with silage when compared to baled hay.[10]

Making forage with the chopper also minimized the risk of rain falling on the cut and cured crop before it could be put in the barn. Rain on cured hay broke it down and leached the nutrients out of the plant material. Farmers could only guess about the weather and hope they could make hay in sunny weather. One farmer who baled his hay reported that he baled on the second day after he mowed, making it a three-day cycle to bale the crop. Experts estimated that it took approximately thirty hours of sunshine to cure hay, but the odds of rainfall at that time of year during the curing period were one in three. Farmers who chopped hay stored it quickly and avoided a damaged crop.[11]

With chopped hay it was also easier to control what cattle ate. When farmers were feeding loose hay, animals could select the leaves, generally the most attractive parts of the plant, and leave the tougher stems behind. Chopped hay on the other hand was comprised of leaves and stems mixed together. William Pfab of Dubuque County testified that cattle wasted less of their feed with chopped hay. "The cows have to eat the tough stems right along with the leaves," he reported, "They can't sort out parts they like best." This was even less of a problem with haylage. A farmer from Cedar County stated that he made grass silage out of his first cutting of hay. "Ensiling makes the stems more tender and palatable," he contended.[12]

Manufacturers and advertisers emphasized the flexibility that farmers gained by using new haymaking machines. The Allis-Chalmers Company developed the Roto-Baler in the 1930s and first marketed it in 1947, claiming that the small round bales were of higher quality than square bales. Round bales preserved more protein because the leaves and stems were rolled together rather than packed. In addition the round, tightly wound bales shed water much better than square bales, which meant that they could be left in the field after the third cutting in late summer or fall. Farmers could turn their cattle out over the winter to consume the bales in the field, a practice that Doyce Miller and his son Lyle used in Clarke County in the early 1960s. Depending on the size of farm, this saved handling hundreds or even thousands of bales. These small round bales were also somewhat difficult for cattle to open, which meant that the animals completely consumed one before opening another, thereby reducing waste.[13]

Costs

Regardless of whether farmers chose to cut operating costs by using balers or choppers, they faced increased capital expenses for the new equipment. During the late 1940s the lowest-cost new equipment on the market was equipment for making loose hay, which cost approximately $543 in 1946. By contrast the new one-man field baler, which automatically tied the bale with twine, cost $2,365. A forage harvester cost $3,486. Farmers who invested in a baler expected to halve the time spent getting hay to the barn. The chopper saved an estimated five-eighths of the labor when compared to making loose hay. The cost of new balers and choppers remained high throughout the 1950s—between $3,000 and $4,000 for a chopper, depending on the size, and as much as $3,500 for a baler, although William Adams of Fayette County purchased a new McCormick-Deering Number 45 baler at the beginning of the 1952 haying season for as little as $1,571.[14]

The issue of farm finances in re-mechanizing the hay harvest demonstrated important distinctions between farmers. Those farmers who farmed large acreages, or who wanted to specialize or to expand, liked the forage harvester since it moved rapidly through the crop and eliminated human handling. Those who continued to practice more mixed farming with a variety of crops and livestock found it less appealing. On the other hand, the baler appealed to farmers with both large and small holdings, because it saved money compared to the older technology and was less expensive than the chopper. Many farmers, in order to reduce costs, purchased used machines and, more important, cooperated with other farmers to purchase a machine together and exchange labor.

Labor exchanges were a popular and less expensive method of conducting the hay harvest. In the prewar period some farmers cooperated to make hay, but the results were mixed. Some participants were unprepared for work and consequently started late or used inferior or dilapidated equipment, while others were ready to go and kept properly maintained machines. Inequalities made exchanging work problematic, but it was a successful practice for many farmers who shared a similar work ethic, managed their farms the same way, or had strong kinship ties. As farmers looked into the postwar world and anticipated continual labor shortages, some of them realized that work exchanges could be beneficial. One farmer stated with a touch of hyperbole, "I guess we neighbors will have to get acquainted with each other and learn to work together."[15]

Work exchanges were especially popular among farmers with smaller acreages who practiced mixed farming rather than dairy or beef production. Two brothers from Marshall County joined several neighbors to form a baling ring and hired a custom baler to do the job for them. In addition to sharing the capital outlay for the machine, these farmers claimed they saved time by dividing the workload and even improved the quality of the hay they harvested. Younger farmers who were just getting started liked the labor exchanges and were more likely to exchange work than established middle-aged farmers who owned more machinery.[16]

Exchanging work during hay harvest also let farmers use family labor more effectively. After the war a higher percentage of Iowa farm youths attended high school, a factor that exacerbated farmers' labor problems during the school year but also meant that children were available to help cultivate corn and make hay during the summer. Around 1954 Charles Havran's oldest son joined the military, which left only one teenaged son at home to help on the farm. When a neighbor family purchased a used baler, the Havran family cooperated, making a team of two men and two boys. They set the baler to make small light bales that were easy for the boys to handle. The Havran family established another exchange when they purchased their own New Holland baler. This time they cooperated with a pair of neighboring brothers. Both families owned balers, so they would mow and bale at the same place at the same time to speed the work and get the hay in quickly to prevent rain damage.[17]

Some farmers took the practice of work exchange one step further and jointly purchased a baler or chopper. This spread out the initial investment in the machine and allowed them to share the operating costs. Farmers accepted the need for cooperative purchasing, especially for father-son ownership agreements. Cooperative ownership among non–family members was potentially more risky, but many farmers sustained successful partnerships. George Lee of Hardin County liked his arrangement with Floyd and Emery Lake. They lived about two miles from each other and invested together in a chopper. When they traded in their chopper for a newer model in 1952, each farmer contributed only $505. Cooperative ownership had another advantage. "What's nice about this arrangement," Lee noted, "is that you get your work done when you want it done," rather than waiting to fit into a custom chopper's schedule. In 1954 Melvin Hansen of Boone County purchased a baler with a second cousin and did not have to invest any money up front. His cousin bought the baler, and Hansen paid off his share over the next couple of years. In 1956, 31 percent of Iowa farmers owned hay balers in partnership, while only 6 percent jointly owned choppers. Farm owners were more likely to have cooperative arrangements than renters, since farm leases were for one year and tenants did not necessarily stay in the same neighborhood from year to year.[18]

Custom baling was a good strategy for farmers with smaller acreages. Balers were expensive, and there were only so many days they could be used, during two or three cuttings of hay in the summer, for baling straw for livestock bedding after threshing or combining, and in a few instances, for baling cornstalks. An Iowa State study of the costs of baling in eastern Iowa during the late 1940s indicated that, in order to make ownership of a baler pay, farmers needed to make at least ninety-three tons of hay per year. Carl and Bertha Peterson's farm account books show the premium they placed on custom baling as well as purchasing baled hay for their 113-acre farm. They hired balers for straw and hay on a regular basis in the 1950s and 1960s. By the late 1950s the pattern of ownership versus hiring balers was clear. Only 36 percent of those farming between 30 and 179 acres owned their own machine, while the rest hired custom balers. Almost half the farmers with over 180 acres owned balers. For most farmers with smaller acreages, custom work made the most sense.[19]

Older farmers approaching retirement also liked the idea of custom baling. This eliminated the heavy work of moving hundreds of bales, each weighing between forty and a hundred pounds depending on the farmer's preference. A 1957 survey of farmers indicated that farmers over the age of fifty exchanged work less often than younger farmers and were more likely to hire custom operators. This was the case for the Petersons. They began farming in 1931 and were at least in their fifties by the 1960s; they relied on custom work.[20]

Farmers who used balers rather than field choppers or a combination of equipment adopted a conservative management strategy. They were less willing or able to expand production or to assume the increased costs and risks of expansion or specialization. For these people with limited goals, baling offered ease of handling and flexibility. This was especially true for farmers who hired custom balers. A farmer from Pocahontas County favored custom baling since he did not have enough hay acres and livestock to justify owning a baler or a chopper.[21]

Hiring custom balers made sense for many people, and others found that purchasing a baler and doing custom work for others could also be a source of income. Custom rate charts were a regular feature in *Wallaces Farmer,* noting average per-bale charges, cost estimates on a per-acre basis, and rates for almost every other farm task from planting to harvesting. Farmers who owned balers and farmed small acreages hired themselves out as custom balers in order to make owning a baler pay. Farm families such as the Petersons provided a ready market for those willing to invest from $1,000 to $1,500 in a baler. A Tama County man reported that the increase in the number of balers was good news for those who wanted to hire balers. "Lots of balers in the country," he stated in 1951. "Maybe I can get it done cheaper this year." For the Petersons custom rates for baling fell by as much as 20 percent from the late 1940s to the early 1950s.

They paid twelve cents per bale in 1948, eleven cents in 1949, and ten cents from 1954 to 1967. When a twenty-two-year-old renter from Shelby County, Vernan Schnack, began farming in 1959, he purchased a new baler for $1,500 to use on his rented farm and to hire out for custom work. "I figured I'd be paying out in custom charges ($150 for three thousand bales) just as much as depreciation on the new machine would be," he explained. "Also, now I can bale when I want to and pick up some money doing custom work." For Schnack, buying a baler gave him independence in his farm operation as well as a method of increasing income.[22]

Chopping Forage

Field chopping was the most effective way for farmers who wanted to expand their acreage or specialize in dairying or beef-cattle feeding. The rise of field chopping was a signal example of how farmers substituted capital for labor. The added expense of the chopper and wagon or wagons made it critical for dairy-farm families who wanted to green-chop their forage to have at least twenty-five milk cows in order to justify the cost. When discussing whether to make hay or grass silage, farm journalists noted that grass silage yielded about 10 percent more feed per acre than grass hay, and this figure grew considerably when rain fell on the cured hay and diminished quality. Silage made economic sense if farmers were in a position to utilize the extra feed value of the silage by increasing their feeding operations.[23]

In the early 1950s farmers used choppers and wagons to haul forage directly to their cattle in lots, bypassing the need to store as much forage and consequently reducing the pressure on pastures. This new system, called green-chop or zero grazing, was a way to use the forage harvester and special wagons to chop hay and move it directly to the feed bunks. By keeping cattle in the feedlot, farmers avoided the traditional pasturage problems of trampling and selective grazing by cattle that chose the most succulent plants and left the rest. Most important, green-chopping was a means of increasing the size of beef and dairy herds. Gail Hemmingson of Plymouth County experimented with the system in 1954 and found that he cut his acreage requirements for feeding his herd, but in 1953 he needed only thirty acres of pasture for fifteen cows with calves. In 1954 he chopped nine acres to keep fifteen cows with calves plus ten heifers. The experiences of a farmer from Woodbury County who rejected zero grazing indicate just how important the practice was to farm expansion and specialization. He acknowledged that he doubled his per-acre carrying capacity for cows but claimed that his small herd did not justify the labor cost of feeding in the lot and concluded that farmers with bigger herds were in the best position to profit from zero grazing. Even though it appeared that cutting acreage requirements for feeding could appeal to farmers with smaller acreages who wanted to increase their herd size, the high cost of the forage made it more economical for farmers with larger acreages who could spread the investment and operating costs over more acres and larger herds.[24]

Intensive management of green-chop or zero grazing allowed significant gains in productivity per farmer and per animal. In 1961 Robert Liston, a Dallas County dairy farmer, boasted that his forty-five dairy cows had not been on pasture in the five years since he began using a green-chop system. He wrote, "I find it easier and cheaper to haul feed to them." Liston began his round of chopping in May and continued into October, cutting two loads for the morning feed and one for the evening, which totaled two hours of work per day. The green-chop and drylot system boosted production and reduced his workload. Drylot feeding also entailed new tasks, however. Liston reported that his biggest problem was manure removal. With the cattle in a lot rather than on a pasture, he moved tons of manure out of his lots to prevent them from becoming quagmires. In 1960 he hauled approximately 190 loads of manure out of the cattle shed and another 100 loads from the feeding area, which meant that he moved manure eight out of every ten days.[25]

In spite of the new work requirements, dairy producers such as Liston were especially interested in this system. In the years after World War II, most farmers in Iowa abandoned dairy production and sold their small herds of approximately half a dozen animals. A minority of farmers invested in dairying in order to cut operating costs and to improve quality. The forage harvester and related equipment, including hay dryers, played a key role in allowing these farmers to stay competitive. Inexpensive forage crops could be substituted for some expensive grain in the dairy ration. In dairy farming, one of the key calculations is the difference between gross income from milk production and the cost of feed. The resulting figure, called income-above-feed cost, is a test of whether a dairy farm can stay in business. As one Fayette County farmer put it, "You have to feed for top production. Yet you have to hold feed costs down if you expect to stay in dairying." He cut costs by replacing grain with silage. Other dairymen echoed this conclusion. Two farmers from northeast Iowa discussed their costs and production with a reporter from *Wallaces Farmer*. Their cows' output averaged from 12,000 to 14,000 pounds of milk per year per cow, but their income-above-feed cost was good, in spite of significant variance in production. The cows that yielded 12,500 pounds per year had an income-above-feed cost of $322 per cow, just $2 less than the cows that produced 14,000 pounds. The key to these dairymen's success was the use of almost twice as much hay and silage as grain or concentrates, which minimized their commitment to high-cost grain and commercial feed.[26]

Storage

The farmers who decided to re-mechanize the hay harvest by baling or chopping their crop faced new storage problems. The large barns on most Iowa farms were built for storage of loose hay, which was bulky and required more space per ton than the densely compressed bales or the chopped hay, which packed tightly in the mow and took up less space per ton. Iowa State experts advised farmers that baled hay weighed three to four times more per cubic foot than loose hay, and that chopped hay was twice as heavy as loose hay. Farmers with overhead or second-storey mows in their barns needed to reinforce their mows in order to prevent structural damage to their largest and most expensive building.[27]

Handling the crop in the barn was also a major problem. Some farmers installed bale conveyors in their barns to reduce the heavy workload of handling bales. Elwood and James Walker of Polk County installed a bale conveyor to solve the labor problem. The bales simply tumbled off the conveyor into a pile in the mow. The Walkers were not concerned about the lack of a tightly packed stack since they believed air circulation was better with the loose pile of bales than with a stack and cut the risk of spontaneous combustion from wet hay. Any lost space was more than offset by the reduced need for one or two men in the barn stacking bales.[28]

Chopped hay posed special problems. Modifying barns by adding hay dryers was one technique to aid the handling of a chopped crop. These dryers forced air through ducts made of wire, sheet metal, or wooden slats in the haymow. Hay dryers accelerated the drying process and enabled farmers to store larger amounts of chopped hay. In 1950 Kenneth Parrett, a farmer who kept a herd of forty-five Holstein cattle on 210 acres in Washington County, began using a dryer in his barn. He believed that bringing hay into the barn with higher moisture content would reduce leaf loss because farmers who field-cured their hay risked losing as much as 15 percent of the total volume of the crop through handling. Keeping more leaves on the hay meant more feed for the investment in labor, fuel, and machinery required to cut, pick up, and haul it to the barn.[29]

Using a hay dryer not only increased the amount of hay farmers could harvest but also improved the quality of the crop. Parrett observed that "barn-cured hay has better color. It's more palatable," which meant that the animals consumed more of it. Eldred Mather, a Floyd County dairy farmer, related his success with a dryer. "The thing I noticed when I started [in 1955] feeding high quality alfalfa that was mow-dried was that it took less protein supplement in the ration to give the same results in butterfat production." Using a hay dryer meant that farmers could cut hay at the proper stage of the growth cycle for optimum nutrition. Promoters of barn drying were careful to point out that it was costly. Only farmers who had large numbers of livestock could make it profitable. As a result hay dryers remained equipment for only a minority of farmers.[30]

Although filling barns with bales or chopped hay comprised the most common method of storing hay, farmers also turned to other ways to deal with hay storage. Using and adapting the old barn might have been less expensive (unless it included a hay dryer) but required a lot of handling to get the crop out of the barn to the livestock. Adaptive reuse of the barn for baled hay or forage saved labor at harvest time, but it did not help at feeding time, a daily chore. If farmers were to realize the full labor savings they craved, they needed either to automate their feeding or to consider new structures for storing the crop that would give them more ease of feeding.

The hay keeper, a cylindrical or rectangular pole building, was an innovation of the 1950s that had the potential to reduce labor costs. Iowa State experts designed the hay keeper as a self-feeder for chopped hay. It was a simple structure of posts (often telephone poles), set in a circular, square, or rectangular pattern, and sheathed in snow fence. It was wider at the bottom than at the top to reduce friction and let gravity pull the hay downward as cattle fed from the bottom. A large cone pointed upward at the bottom forced the hay toward the edges and into the mangers for feeding. This feature made the structure attractive to farmers who wanted to expand and simultaneously cut labor costs. "The self-feeding feature is what sells me on haykeepers," noted a dairy farmer from northeast Iowa. Leon Wengert of Story County constructed a rectangular hay keeper for beef and dairy cattle. This structure—fifty feet long, twelve feet wide, and twenty feet high—housed his first cutting of hay, since it could be stored in the hay keeper with less risk of spontaneous combustion than in a barn. It also helped cut labor costs at chore time. Extension agents offered plans for fifty- and sixty-ton models, with costs ranging from eight hundred to eleven hundred dollars, depending on size as well as local costs for materials and labor. Hay keepers could also be built with central ventilator shafts equipped with a large fan and motor to speed the drying of the hay. This feature increased the cost of the building but reduced the amount of time the crop needed to dry in the field, since more drying occurred in storage.[31]

In contrast to the low cost of hay keepers, upright silos were a costly solution to the problem of making high-quality feed. The high first-cost of upright cylindrical silos generally meant they were reserved for the corn-silage crop rather than the hay crop. Farmers, researchers, and extension professionals learned that corn silage had higher nutritive value than haylage, which meant that it was better to use the high-cost silo for the highest-quality feed. Some farmers even used their silos for storing high-moisture corn that had been harvested by combines or picker-shellers, popular in the 1950s, rather than by the older corn pickers, which harvested the entire ear instead of just the grain. With high-moisture corn and silage in the silos, farmers who chopped searched for other places to store haylage.

Farmers reduced haylage costs by making trench silos, a decades-old but still popular and inexpensive storage method of keeping corn silage. After leveling and scraping a strip of earth measuring approximately twenty-four by seventy-five feet or large enough to accommodate the crop, builders constructed a fence along the edge and lined it with Kraft paper to keep oxygen out and to prevent spoilage. Then they dumped the chopped hay onto the scraped earth, leveling and packing it as they piled it deeper and finally covered it. As farmers needed feed, they either used a tractor loader to haul silage to livestock or let cattle eat directly from the silo by controlling where they ate with electric fence.[32]

Bale storage was easier and less expensive than storing haylage or chopped hay. Some farmers built low-cost hay sheds to store bales, or they even stored bales out in the open. One Greene County farmer built an "umbrella shed" measuring thirty-one by fifty feet, to store thirty-two hundred bales. A line of five center posts supported the roof, with no side or end walls to impede stacking and loading. The owner, Paul Williams, stated, "The quickest and easiest way of unloading and storing hay was what we were after." Williams simply pulled up next to the stack without worries of hitting walls or supporting members with his tractor or hayrack. Outside storage eased the labor of unloading and stacking, but there was some loss through spoilage along the top of the stack. Some farmers accepted this small loss as a preferable alternative to a large investment in a new hay shed. Removing bales from the end of the stack rather than the top minimized the inevitable loss associated with stack storage and ensured quality feed for livestock. Then, also, the Allis-Chalmers Roto-Baler made bales that were well suited to outdoor storage and self-feeding for beef cattle. Farmers who used these machines further reduced the need to store hay. Regardless of how farmers solved their storage problems associated with the new haymaking machines, they envisioned a future free of the intense and time-consuming work of putting up loose hay in the traditional way of previous generations.[33]

Hands-off Haying

The number of balers and choppers on Iowa farms reflected Iowa farmers' interest and confidence in the new technology and their ability to adapt work routines, finances, and buildings to the new balers and choppers. In early 1962 there were 49,177 balers and 12,774 forage harvesters on Iowa farms, and in 1970 there were 48,052 balers and 15,159 forage harvesters. Although the number of balers had declined by 1970, the proportion of balers to the number of farms was actually higher, since there were fewer farms in 1970 than in 1960. In 1961 there were 177,172 farms in Iowa, with a ratio of one baler for every 3.6 farms. In 1970 there were only 135,264 farms, with a ratio of one baler for every 2.8 farms. The increase in the number of forage harvesters was even more striking in comparison to the changing number of farms. In 1961 the ratio was one harvester to every 13.8 farms, and in 1970 it was one harvester to every 8.9 farms. Farmers rushed to make hay the modern way.[34]

In 1972 balers were essential on all Iowa farms that included livestock. Not all farmers owned balers, but almost all farmers used them. Farmers with small acreages frequently hired custom balers and paid a flat fee per bale. Other farmers, with both large and small holdings, purchased balers in cooperation with friends, neighbors, or family members to minimize the initial outlay and to share operating costs. These farmers were less likely to enlarge their livestock operations or to specialize in beef or dairy animals. They kept a mix of animals and crops while they continued traditional work exchanges. Even farmers who specialized in one particular area valued hay balers. It was cheaper for them to bale some hay to maintain flexibility in their feeding programs than to feed only green-chop hay or haylage.

Only a minority of Iowa farmers purchased forage harvesters, although these machines were essential on larger and more specialized farms. They were very expensive, but they could also be used for harvesting hay and corn silage, which spread their cost over more of the farm operation. Chopping hay was quicker than baling, which made it possible to harvest more forage and to feed more livestock with better-quality feed. Using a forage harvester was an important strategy to beat the cost-price squeeze by offsetting declining commodity prices with improved-quality feed and more production.

In 1967 farmers learned about a new baler that produced a giant bale, which allowed producers to gain even more labor savings and which literally increased the scale of production. The baler produced a cylindrical hay bale of up to seven hundred pounds, with the potential of making one that weighed a thousand pounds. Iowa State engineers developed the machine with the goal of mechanizing hay handling completely. One operator with a tractor could handle these huge bales, transporting and feeding them to livestock without ever touching the crop at all. Farmers who used the new baler could avoid the strenuous and repetitive lifting they were accustomed to doing themselves or that they hired others to do for them. Wesley Buchele, a mechanical engineer at Iowa State, summed up the purpose of the new machine. "The giant bale is designed for mechanical handling," he stated, "while conventional bales are designed for human handling." It is difficult to know if farmers perceived this development as a sign of things to come or something so far-fetched and ridiculous that it would never be practical. If the postwar trend of substituting expensive machinery for scarce and expensive labor continued, farmers may have looked at the bale and glimpsed a future of completely mechanized hay handling.[35]

Seven

From Threshing Machine to Combine

Sit-Down HARVEST
RIDE—AND WHISTLE WHILE YOU WORK;
FAMILY HARVEST INSTEAD OF THRESHING GANG

The man operating an All-Crop Harvester stopped near the fence to talk. "Look at my shirt," he said, "it's dry! I thresh sitting down."

A dry shirt is only a symbol of the freedom and independence an All-Crop Harvester brings you and your family. DIRECT harvesting that frees you from the backaches of shocking . . . the dust and sweat of threshing . . . from big customs bills—and frees Mother from cooking for extra men.

> —Allis-Chalmers All-Crop Harvester advertisement,
> *Successful Farming,* March 1940

When this Allis-Chalmers advertisement appeared in the spring of 1940, combines were relatively scarce on Iowa farms. Few Iowa farmers threshed their grain without soaking their shirts with sweat. The practice of harvesting and threshing with separate tools and at different times was the prevalent method of making small-grain crops such as oats. Farmers used tractor- and horse-drawn binders to harvest the grain, set up the bundles into shocks by hand, let the shocks dry in the field for a week or so, and then hauled the bundles to tractor-powered threshing machines in the barnyard. There, family members and neighbors gathered to feed bundles to the threshing machine, scoop the grain from wagons into the bins and granaries, and stack or bale the straw for use throughout the following year.

Between 1940 and 1972 most Iowa farmers turned to combines for harvesting their small grains and rejected the separate tasks of binding and threshing that had prevailed since the 1880s. When the implement industry developed a machine suited for the small farms of the Midwest, farmers began to replace the old tools with the new. The decision to use combines was based in part on the appeal of "sit-down" and "dry shirt" ease, but farmers found more compelling reasons for abandoning their existing set of tools. Farmers in Iowa used combines during these years because they could afford to hire or purchase the machines, the technical obstacles were minimal, families enjoyed the work and leisure benefits of increased mechanization, and the combine helped facilitate the transition from oats to soybeans that began in the mid-twentieth century.

Combines were not new inventions in 1940, but they were relatively new to the Corn Belt. Inventor Hiram Moore and financier John Hascall had developed a workable combine by the 1850s, sparking a chain of developments that led to the general adoption of combines in the arid regions of the United States by the 1920s. These machines were tractor-drawn, powered by either an auxiliary engine or the tractor's power-take-off, cutting a swath of eight feet or more through the massive fields of the Great Plains, California, and Washington. A handful of Iowa farmers used these combines on their farms, and they invariably performed custom work for other area farmers also.[1]

Combines for Midwestern Farms

In 1930 inventors developed a prototype combine that farmers with the smaller acreages of the humid Midwest could afford to own. Both John Deere and McCormick-Deering produced tractor-drawn models that they planned to sell in the region, but these had ten-foot and eight-foot cuts, respectively, making them too large to be efficient for mixed farming. That year *Farm Implement News*, a journal for dealers and manufacturers, printed three stories about a "baby" combine that might work well on midwestern farms. This machine was pulled behind a tractor, and it cut a five-foot swath, which made the machine practical for a region where 160-acre farms were the average. The threshing cylinder was also five feet long, which maximized operating speed. In addition to threshing almost any crop with a cylinder of wire brushes, the machine had low horsepower requirements, making it suitable for farmers with older or smaller tractors. Allis-Chalmers purchased the license to develop and manufacture the Fleming-Hall machine. The company eventually scrapped the wire-brush design, because during testing the wire bristles broke off and mixed with the grain, which posed a hazard to livestock.[2]

Agricultural engineers during the 1930s conducted extensive tests on small pull-type combines, analyzing grain loss, speed, and performance in varying conditions. A 1936 study concluded that machine adjustments and crop conditions were more important in determining grain loss and quality of harvesting than the size or type of machine. The study authors also noted that the small size, light weight, and pneumatic tires were all important aspects to consider in a region with relatively small or irregularly shaped fields. One aspect of Corn Belt combines was their ability to handle midwestern conditions, especially green weeds and "rank," or thick straw. Since the cutter bar was just as wide as the threshing cylinder, the machine handled heavy crops without clogging. An engineer from the J. I. Case Company recorded strong objections to the combines; most notably that the small-sized tractors commonly used in the Midwest did not have enough power to operate the machines. According to the Case engineer, "typical" farmers would be unable to operate "within the speed limits essential to efficient threshing, separating, and cleaning." But engineers generally favored the new combine in spite of concerns about harvesting losses and fears that the small machines were easily overworked.[3]

As the engineers were testing the "baby" combines in the mid-1930s, a small number of Iowa farmers used the larger models, either purchasing one or hiring a custom operator. E. M. Brubaker of Prairie City, proprietor of an implement dealership during the 1930s, had his sons custom harvest with two large Rumley combines, but farmers who used combines were still in the minority. Even after Allis-Chalmers introduced the All-Crop combine (their production version of the Fleming-Hall baby combine) in 1935, binders were still more popular than combines, outselling them ten to one. By the end of the decade, however, sales of combines had overtaken sales of binders or threshers. In 1939 combines outsold binders by two to one and outsold threshing machines by eleven to one. Doyle Brubaker, who set up and adjusted combines for his father at the family implement dealership in Prairie City, noted that in 1940 "you couldn't give a binder away; the market was gone."[4]

Purchasing a combine made good economic sense for a farm family. Allis-Chalmers Company introduced the Model 40 All-Crop in 1938, their smallest pull-type combine with a forty-inch cut, priced at $345, the same as a power-take-off binder. When faced with the decision to replace a worn machine or invest in a combine, farmers could make a clear choice; purchase the older technology that was reliable and familiar or obtain a new machine with the potential for cost savings. Running a combine through the field consumed less fuel than operating a binder, hauling the bundles, and powering the threshing machine. Furthermore, families saved money on twine and the cost of feeding a neighborhood threshing crew.

Salesmen stressed the potential cost savings as they met farmers and spread the news about the new combines. Allis-Chalmers representatives visited local dealers and hosted suppers for area farmers, following up the supper with a sales meeting. Doyle Brubaker related the selling technique used by one representative in Iowa during the late 1930s. The local dealer provided the salesman with information about leading farmers in the area and then called on one of them at the evening meeting. After he asked the farmer about how much grain he harvested and what expenses he incurred to harvest the crop, the representative brought home the point by stating, "you know, every two years you're buying an All-Crop Harvester and you don't have one." The comparison of the costs associated with binding and threshing made purchasing a new combine look more reasonable.[5]

By the late 1940s and 1950s, farmers who did not want to purchase new machines could purchase used models, thus easing the transition to the new technology. The Robinson family from the western part of the state bought a used Massey-Harris combine and kept it for at least two years before they purchased a new machine of the same make. In 1955 the Holm family of Eagle Grove purchased a 120-acre farm in Lincoln Township, Madison County, and invested $1,591 in equipment that year to commence farming. Their largest expenses included $400 for a 1947 tractor and mounted cultivator, and $350 for a six-year-old International combine. The cash saving on the used machines was significant and minimized capital expenses. For example, the Holm family's used machine cost only a few hundred dollars in 1955, while Elmer and Darlene Meyer bought a new International combine in 1954 for $1,750. Frequently farmers who purchased combines used them for custom work, which also offset the cost.[6]

Farm writers compared the costs of combine ownership versus rental. In 1952 the staff at *Wallaces Farmer* concluded that farmers who harvested less than 65 acres should hire rather than purchase a combine. They calculated that the harvest on a 173-acre farm with 40 acres of oats and another 25 acres in soybeans would cost $292.50 (that is, $4.50 per acre) for custom or hired work. By contrast owners would pay about $100 per year in depreciation on a $1,500 combine, with another $110 for housing, upkeep of, and interest on the machine. Adding $2.00 an hour for tractor and operator covering 1.5 acres per hour, operating costs equaled $1.35 per acre or $87.75. Added to the overhead cost of $210.00, the total harvest cost would be $297.75, slightly more than custom work. The authors concluded that farmers with less than 65 acres were wiser to hire their combining.[7] Two years later *Wallaces Farmer* published an almost identical article, this time with calculations based on 70 acres to combine, concluding that farmers who raised between 60 and 70 acres could make a combine pay, especially if they did custom work or bought a used machine. Writers from a competing magazine, *Successful Farming*, recommended that farmers who had less than 80 acres of their own and could not arrange to have another 20 acres to harvest should not purchase a new combine.[8]

The experience of Rudolf Schipull of Eagle Grove illustrates the financial considerations of combine technology. Schipull kept extensive records of his farming operation from 1939 until he retired in 1952. When he commenced record keeping in 1939, Schipull conducted an inventory of his equipment. Among his machines he owned a 1934 McCormick-Deering binder valued at $223 that he used to harvest his oats. Each year until Schipull sold his equipment in 1952, he recorded the expenses for equipment repair, twine, threshing machine costs, and custom combining charges for a variety of crops including soybeans, flax, and clover seed.[9]

Schipull's records detail the significant expenses associated with maintaining a binder-and-threshing outfit. From 1939 to 1950 expenses for binder and separator repair ranged from a low of $1.47 in 1940 for binder sections (cutting surfaces) to a high of $28.25 in 1943. That year Schipull spent $17.40 for a new binder platform canvas (to convey the grain across the machine), $1.64 for twenty-five binder sections, and $1.28 for a new separator belt. The largest equipment expense he incurred was to maintain the tractor he used to provide power for his separator. He made the final payment of $400.00 in 1939 and spent an average of $77.29 annually for maintenance from 1939 to 1950, when he bought a combine. Schipull's records show that by 1950 the threshing tractor had become a problem. From 1939 through 1945 the average yearly cost to repair that tractor was $43.42; the costs from 1946 to 1950 averaged $124.72 annually. During the last five-year period of use, Schipull's repairs included replacing transmission bearings, two valves, crankcase and camshaft bearings, piston rings, a connecting rod, and numerous other adjustments.

It is impossible to know if the mounting expense of the threshing tractor prompted Schipull to invest in a combine, but a comparison of the expenses for the old technology with the new technology suggests that, when faced with sizable repair bills, he found the combine promised equal or lower costs. Schipull paid $200.00 for one half-share of a combine after spending almost $125.00 a year for over five years on repairs to the threshing tractor. In addition to tractor repair, he incurred annual expenses of almost $32.00 for twine and close to $8.00 for repairs to an aging binder-and-threshing machine, altogether meaning it cost him approximately $165.00 to harvest his oats. When the average annual expense of $83.30 for custom-combining soybeans from 1939 to 1951 is added to this figure, one sees that Schipull spent $248.30 every year to conduct his harvest, which more than offset the cost of his share in the combine.[10]

In addition to machinery costs, repairs, and supplies at threshing time, there were also costs associated with laborers. Threshers' meals were regular expenses that fluctuated depending on how large a crop a farmer planted. In 1947 Rudolf Schipull purchased thirty-two meals for threshers during two days in August, spending a total of $21.42. Joseph Ludwig of Winneshiek County regularly purchased beer and "pop" for his threshers. In 1948 he spent $8.50 for two cases of beer and one case of soda. While many families exchanged labor, which minimized cash expenses, families sometimes hired help for threshing. Rudolf Schipull hired laborers to help with his threshing, increasing the cash costs of an already expensive operation.[11]

By the middle decades of the century, threshermen such as Schipull were contending with a new type of threshing ring, which influenced farmers' decisions to purchase or hire a combine. Once farmers owned their own tractors, which a majority of Iowa farmers did by 1940, they were liberated from the need to hire a steam engine for threshing. Instead of the large rings that used steam power, farmers who threshed in the late 1930s and 1940s frequently used their own tractors and smaller threshing machines, and the threshing rings comprised fewer members. In 1945 Herb Swaggart, a thresherman from Hardin County, recalled his days of operating large threshing machines powered by steam engines in the early 1900s. Swaggart noted that previously he had conducted two August threshing runs, one with a ring of fourteen or fifteen jobs and the second with sixteen jobs. The sixteen-family ring required the labor of as many as twenty-two men and their families. By the mid-1940s, threshing rings like the ones Swaggart remembered were the exception rather than the rule. Farmers and journalists perceived a significant difference between the threshing of an earlier era and the threshing conducted on farms during the late 1930s and 1940s. With families using small threshing machines that handled six or eight jobs, farmers worked with fewer neighbors and relatives.[12]

Johnnie Westphalen's threshing experiences in western Iowa illustrate how threshing was changing. In the early 1900s Westphalen's family participated in a twenty-seven-member ring that covered parts of southern Audubon and northern Cass counties. Like Swaggart's family, these families threshed together for almost the entire month of August. When Westphalen and his wife, Marjorie, began farming on a rented thirty-acre farm in Audubon County in 1939, he started a threshing operation to supplement the income from his small farm. However, the rig and the threshing ring Westphalen used were very different from the tools and organization his parents had known. The younger Westphalen used a twenty-two-inch McCormick-Deering threshing machine and purchased a used John Deere Model D with money he borrowed from his uncle. Westphalen utilized his "little rig" to thresh for five or six neighbors in his family neighborhood as well as for his in-laws in the Buck Creek neighborhood. He found a niche in the new rural landscape where tractors democratized threshing-machine ownership.[13]

As the older threshermen died or retired, their younger counterparts could not or would not sustain interest in the older technology. Schipull, an established farmer, probably continued threshing because his equipment was unencumbered by debt. Once he paid off his threshing tractor, he only had to worry about upkeep, and when that became too onerous, he switched to combining. One Cass County farmer returned from military service in Korea and participated in one of these small rings for a season or two. He then purchased a used combine and exchanged tradition for the advantages of harvesting on his own schedule.[14]

Combine Operations

Once farmers decided to use a combine, they had to learn the best operating techniques, since combine operation involved more than just hitching, starting the tractor, and heading into the field. When was the optimum time to harvest with a threshing machine? Each stand of grain ripened unevenly, depending on field conditions and the variety of seed planted. How could combine operators ensure that the oats would be ripe enough, without waiting too long and risking grain loss at the cutter bar due to shattering? To prevent threshing loss from cracked seed or from underthreshing, farmers needed to adjust the distance between the cylinder and concave and to set the cylinder at the proper speed for the type of crop. Similarly, the flow of air had to be adjusted to match crop conditions and crop type and capture all of the threshed grain.

These issues, if not properly addressed, could make the difference between profit and loss, the technology's success in meeting the farmer's needs. Farmer adaptations, advice from farm periodicals, and instructions from the manufacturer and sales staff all helped farmers accept the combine. But the willingness of farmers to experiment, on the faith that sooner or later the new machines would help them harvest with greater financial and labor savings, was the most influential factor in their acceptance of the combine.

Although farmers believed that combines could save money and labor when compared to binders and threshing machines, the transition to combines involved technical obstacles. A perennial debate about combine use in Iowa concerned the proper time for harvesting. According to writers in *Wallaces Farmer*, "One problem in connection with the spread of combines is teaching the owner to avoid harvesting too soon." As long as farmers used binders and then threshed their grain a week or ten days later, farmers cut grain when it was only partially ripe. "With the new machines," the journalists cautioned, "it is advisable to hold off an extra week or ten days past the binder stage." If the farmers waited, the grain ripened in the stand until it was fully developed and could be stored in the bin at the proper moisture content. If the stand was in danger of lodging (falling over) or was especially weedy, then the farmer could windrow it and let it lie in the field to dry. Articles contained information on letting the crop cure in the windrow for a few days to get it dry enough so that the combine could separate the oats from weed material (known as trash). Moist trash mixed with the grain could cause the grain to spoil in storage. Farmers needed to make sure that the crop they stored was at 14 percent moisture content to prevent deterioration in the bin.[15]

The standing-versus-windrow debate continued in the weeks leading up to the 1942 harvest season. One writer noted that farmers disagreed about whether it was preferable to combine standing grain or to windrow it before combining. He "had almost come to the conclusion that windrowing was practically imperative with oats, perhaps as much so as for flax," until he spoke with Glen Blanchfield of Lake City who regularly combined standing oats. While not taking sides, the author counseled farmers who used both methods. To ensure a ripe crop, farmers who combined standing grain needed to wait at least ten days after the time when they would normally start binding. The author noted that "ripe oats does not go to pieces as badly as we thought it would." Windrowing required stubble just a little higher than with binder operations in order to keep the windrow far enough above the ground to allow the grain to dry without bending the stubble under the weight of the windrow. If the cut grain touched the ground it would not dry properly and could even mold in the field.[16]

By the end of the 1940s, admonitions to windrow oats before combining had become an annual feature in *Wallaces Farmer*. In 1946 a Marshall County farmer, Merle Stansfield, observed, "There are a lot of fellows who just can't wait. But if you want good, heavy berries, you should let oats ripen on the stem." Stansfield used a windrower because he believed that weeds could dry out in the windrow and prevent clogging the combine. By contrast a Hardin County farmer argued that windrowing actually increased the likelihood that trash would get into the grain bin. The regularity of advice to windrow suggests that the process of combine adoption involved trial and error based on local conditions and operator preferences. Regardless of the method, all farmers concluded that the oats needed to be ripe before harvesting.[17]

Wallaces Farmer repeated the same theme of cutting only ripe grain, although authors declared that windrow harvesting was the preferred method. In 1949 the editors came out in favor of windrowing, contending that it helped ensure grain quality. The next year a similar article titled "Windrowed Grain Keeps Better" echoed this finding. In spite of the editor's strong case for windrowing, a 1951 poll found that three out of five farmers who used a combine cut their grain in the stand rather than windrowing it first. Dale Hull, an extension engineer at Iowa State, traveled through Iowa at harvest time. He observed that farmers who sold grain for cash tended to windrow oats. By 1959 it was clear that farmers in the northern part of the state preferred to windrow while those in southern Iowa combined standing grain. The actual practices of farmers indicate that they used machines the way they believed suited their operation rather than heeding the advice of the journalists.[18]

The debate about whether to harvest standing or windrowed grain illustrates that farmers had to select appropriate techniques to use the new technology, but farmers also made their own technological adjustments to solve problems or cut costs. Farmers who preferred windrowing had to get the grain cut and laid in a windrow, which required an extra trip through the field with special equipment. Faced with purchasing a windrower or hiring one at an extra charge, farmers found their own solutions. Some farmers simply used their mowing machines to harvest and then raked the cut grain into a windrow, although one writer noted that "ordinary mowing and side-delivering can sometimes make a terrible rope [of cut stems and grain] to handle." Other farmers converted their binders to windrowers, blending their existing tools that were already paid for with new tools to get in the crop. The easiest way to windrow was to disable the tying mechanism on the binder, which allowed the cut grain simply to slide off the machine and onto the stubble. In the early 1940s Loyd Reidesel of Carroll County made more elaborate modifications to his binder. He removed a section of the binder platform, leaving a drop hole for the cut grain to land gently on the stubble. By the late 1940s Iowa State experts estimated that farmers used eight times as many binders converted to windrowers as they did factory-made windrowers. Charles and Minnie Havran of Benton County purchased a new Allis-Chalmers All-Crop 60 in July 1950 and used a modified binder for two years before they bought a used nine-foot Case windrower. Farmers who combined soybeans also modified their machines. Plant material occasionally jammed in the straw rack of the All-Crop combine when the vines were rank. One farmer's solution was to take pieces of wooden fruit crates and secure the wooden slats over the straw rack to allow the plant material to move over the rack and out of the combine while the beans fell through.[19]

There were other challenges in using combines that threatened their adoption. In July 1947 Ray Gribben of Dallas County hired two neighbors to combine his wheat. One of the men spent the first day adjusting his John Deere 12-A combine for the proper threshing speed, while the other successfully used a McCormick-Deering. The harvest was on schedule during the second day, and on the third day the harvesters wanted to finish, which meant that they continued to harvest until 9:30 at night. In that day's diary entry Gribben recorded how impressed he was with his harvesters' work ethic, but the next day he tested the portion of the crop harvested after dark and found it was too moist to store. He concluded that the grain had absorbed moisture from the night air. In spite of his concerns about the high moisture content of the evening harvest, Gribben did not want to go back to threshing. Instead, he resolved that, if he had wheat to harvest next year, "I'll not permit [combine] harvesting late in evening or early morning." Gribben recognized that there were risks associated with combining and needed to be assertive about minimizing or reducing problems.[20]

The straw problem was another challenge for farmers that forced them to find new solutions to unanticipated problems. Farm families who were raising livestock needed the straw, a by-product of threshing, for bedding and manure removal, but the combine left the straw in the field. Farmers who used threshing machines could blow the straw into the barn or use a straw stacker to create a stack close to the livestock. Throughout late summer and into the fall farmers then used a stationary press to bale the straw for easy handling. According to Keith Robinson of Cass County, "My dad was a little later getting a combine than others because he liked that straw pile . . . he liked that in the winter time for the hogs to get in there." The straw was cheap shelter for hogs, and the animals consumed any unthreshed grain in the pile. Farmers in the Irwin area of Shelby County reported in 1940 that, compared to threshing, combines wasted the straw.[21]

The solution to the straw problem was the pickup baler, a parallel development in harvest technology. In 1941 an observer noted that farmers were "getting around the straw problem by calling in a custom baling outfit to follow the combine." Once families baled their straw, they could pick it up from the field with family labor and haul it to the barn or hog house for storage, using the straw as needed, just as they had done with the straw stack when they threshed. Ray Gribben hired custom balers to bale his wheat straw after his shaky combining start in 1947. After purchasing their combine in 1949, the Anderson family of Audubon County hired a custom baler to bale enough of the straw to get them through the next year. A 1951 farm poll showed that 83 percent of respondents baled their straw. Even though baling was the most common method of saving straw, farmers employed other techniques also. A small minority of farmers reported using a field chopper to cut straw and blow it into a barn or other outbuilding, or into temporary storage made from snow fences. Some farmers said they used hay-loaders or other hay tools to pick up the straw. Most farmers found that their straw problem, created by the decision to use new combine technology, could be solved by other new technology.[22]

Family Labor—From Threshing Ring to Combine

Families who used combines not only changed the type of work performed on Iowa farms by supplanting threshing with combining and straw baling, they also cut the labor needed to harvest the crop and thus altered family work patterns. Families in threshing rings required the labor of almost every family member, including children, for harvesting and threshing the crop. Darlene Meyer of Adair County recalled that her entire family worked in the evenings to shock the grain that the men had cut during the day. Young children hauled water to workers in the field. Older children helped drive horses or tractors on the racks for the bundle pitchers. Teenagers pitched bundles and scooped the threshed grain from the wagons into the bins. Keith Robinson's father and a neighbor drove the tractor and binder while the two Robinson boys shocked the grain. But the need for family labor diminished as combines became popular. With a combine, one or two people could handle the crop from the field to the bin. One Audubon County farmer drove the tractor to pull the combine while his teenage son used another tractor to haul the grain in from the field and then scoop it into the bins, completing the harvest of forty-five acres with just two field-workers.[23]

Combining also reduced the labor of the women who had prepared the large meals for the threshing crews. Some women were happy to be free of the burden of feeding threshers, even the reduced crews of the late 1930s and 1940s. In many cases women were responsible for feeding the crew an evening supper in addition to the noon dinner. Both midday and evening meals were extensive affairs, featuring several kinds of meats, vegetable dishes, pickles and preserves, and bread as well as pies and cakes. In 1940 a contributor to *Wallaces Farmer* wrote a long article urging the "streamlining" of threshing meals. The author, "Mrs. Leslie," noted that the older generation expected the elaborate meals, but that farming and rural life were changing, with different tools and conveniences such as the automobile. She suggested that the men could drive to town and have a good meal at a café where the facilities existed to feed large numbers of people. Furthermore extensive home-cooked meals wasted money and were actually too much for the men. "Men haven't iron clad stomachs," she wrote, "and many a man who begins to look a bit wan around the gills and feels weak in the middle by the time the threshing is half over could come thru a season of ordinary 'vittles' hale and hearty."[24]

In 1942 the magazine backed away from "Mrs. Leslie's" reformist position. Wartime conservation measures of gas and tire rationing provided justifications for feeding the men on the farm rather than sending them into town. In this instance the increased demands of wartime production, the need to make do with existing equipment, and the attending social customs of threshing took precedence over calls for reform. Relief from threshers' dinners arrived later, with the combine.[25]

The diminished need to feed threshers was one of the most significant changes associated with combine use. "With the decrease in large threshing rings and the like," a writer observed in a *Wallaces Farmer* article from 1950, "dinners for big crews of men are becoming less and less common."

When Darlene and Elmer Meyer of rural Bridgewater began farming after their marriage in 1942, Darlene was up at five o'clock in the morning to kill and butcher the five chickens she needed for that day's meal. When the Meyers purchased their first combine in 1954, a new International Model 64, Darlene no longer prepared special meals in large quantities. While combining did not necessarily change the daily bill of fare for rural dinners, women no longer fed a neighborhood crew. Instead they focused on providing for their own family and, on some farms, assisting with the fieldwork.[26]

Women sometimes played a larger role in fieldwork in the Midwest, as hired men left the farm and more farmers turned to machine power. The decline in communal labor put a premium on women's labor in some families. A photograph of a family in an oat wagon next to a combine in the August 7, 1945, issue of *Wallaces Farmer* included a caption describing the oat harvest as a "family enterprise" for the Gurnett family of Linn County. Mrs. Gurnett used the tractor to haul the grain from the combine to the bin. On the Meyer farm Darlene unloaded the oats and Elmer operated the combine and hauled the grain. For most women, however, the postwar period generally marked a decline in the importance of women's work on the farm.[27]

As farm families with combines compressed the grain-harvest season from two, three, or four weeks down to two or three days, families had more time for recreation. In a promotional film for John Deere 11-A and 12-A combines, the fictional Sheppard family considered purchasing a combine to help them perform their harvest more quickly and at lower cost as well as for the added benefits of reduced work and increased time for family recreation. After the Sheppard family purchased their combine and successfully completed the harvest, the father, Fred Sheppard, observed, "If it hadn't been for that John Deere combine, we'd have had a month of hard and sweaty work setting up bundles and threshing." The son replied that it only took four days to harvest their sixty acres of oats at half the cost of binding and threshing. After acknowledging that "Mother" would enjoy a vacation as well as the new electric refrigerator they planned to purchase with the savings, the family departed for a fishing trip with relatives. As the family drove away, they stopped to wave at their neighbors who were threshing just down the road. The film's narrator concluded the dramatization by stating, "You and your family can take a vacation next year if you shorten your harvest with a John Deere combine."[28]

Like the fictional Sheppard family, real Iowa families prepared for a new type of lifestyle. If farm families could persuade or hire someone to milk the cows and take care of the livestock, they could justify a vacation of a few days or even a week or more during the time they previously needed for threshing. A photo essay entitled "How to Take Eight on Your Vacation" from *Wallaces Farmer* depicted one family's preparations for a 1950 vacation to Minnesota for fishing and then across Canada to Seattle and back home. The Gaines family of Linn County purchased an old school-bus and renovated it with numerous conveniences, including bunk beds, electric lights, window screens, a gas refrigerator, and water tanks. The final photo caption in the essay stated, "This is how they'll start out when combining ends and the bus is ready." One Audubon County family decided to take a fishing vacation in Minnesota after they purchased a new John Deere 12-A combine in 1949. In addition to saving harvest labor and time, the income from custom work actually paid for the vacation. Families who purchased combines, then, experienced a change in their lifestyles.[29]

Tenants and landlords also experienced changes in work patterns and organization when they used combines. Traditional threshing arrangements included labor and cost-sharing agreements. Both tenants and landlords did their own binding and split the threshing bill and twine costs. But combining defied the traditional harvesting logic. If an owner who did not own a combine hired a tenant who did own one, the reduced labor requirements and expense of hiring the machine meant that splitting machine costs shifted a burden to landlords, since they now paid cash for what they formerly received as part of the tenant contract. Farmers searched for workable arrangements, which included splitting the cost in half between landlord and tenant, minus the landlord's share of fuel.[30]

Combines, Soybeans, and Changes in Land Use

The decision to purchase a combine played an important part in the transition to soybean culture in Iowa. Soybeans gained rapid acceptance in Iowa after World War II. Journalists debated the best way to use combines in small grains. The first soybean crops were most valuable as forage. Before 1940 farmers used as much as two-thirds of Iowa's small soybean crop for hay, but wartime demands for edible oils for humans and high-protein meal for livestock spurred an increase in acreage for the crop. Farmers invested in cooperative soybean mills to process the crop in Manly, Sheldon, Eagle Grove, and numerous other communities, creating an infrastructure to deal with the surge in acreage from 1.25 million acres (for hay and beans) in 1939 to over 2.2 million in 1943. The wartime boom in soybean production helped create a high demand for combines. In 1942 some areas of the state had enough combines to harvest the bean crop, particularly the grain-livestock regions of the west and south, but in the rest of the state, where soybeans were most commonly grown, the numbers of acres per combine were high. In Emmet County, each combine operator needed to harvest from 200 to 422 soybean acres. Some observers predicted that after the war soybean acreage would fall by half (from wartime highs of over 2 million acres), but in fact in 1954 Iowa farmers matched their wartime production, and in 1961 they raised over 3.3 million acres, and this gave Iowa farmers an additional reason to purchase or hire combines.[31]

Combines were the harvesting machine of choice for soybeans, the region's newest crop, even before the machines were popular for harvesting small grains. A *Wallaces Farmer* article from 1943 noted that farmers accepted combines as the best way to harvest soybeans, even as they continued to use binders and threshing machines for their small grains. Farm records show the coexistence of two types of technology for harvesting and threshing. Rudolf Schipull hired combines for flax, clover, and beans in the 1940s while continuing to bind and thresh his oats. Asa T. Meelchryst threshed oats on his Guthrie County farm in the summer of 1943 but spent twenty-six dollars to hire a combine for harvesting his beans that fall.[32]

One of the biggest challenges faced by combine operators harvesting soybeans was the chance of cracking or splitting the beans. Soybeans, like any seed crop, begin to deteriorate as soon as the outer hull cracks, which results in storage problems if too many split or cracked beans are present in the bin. As early as 1942 farm journalists encouraged farmers to "Thresh Beans with Care" and recommended that operators should "reduce the cylinder speed or there will be too many 'cracks' or 'splits.'" If there is too much chopped plant material over the sieve in the rear part of the machine, the beans "will be lost or will be returned to the cylinder and cracked."[33]

Just as the editors repeated lessons of the oat harvest, they also reiterated the message of making proper adjustments for soybean harvesting. Recommendations to check the cylinder speed, the distance between concave and cylinder, and the air blast were annual caveats. In addition to checking the grain in the tank for quality, farmers needed to examine the stubble to make sure the cutter bar was set low enough to get the low-lying bean pods. In 1951 an expert reported that "you can expect a 10 per cent loss when you combine soybeans . . . [and] tests have shown losses as high as one-third." Constant adjustment and oversight was the remedy for poor combining. One expert noted that careful operation could keep combining losses to less than 5 percent and encouraged farmers to analyze their operation. By counting the number of beans on the ground in square foot blocks, farmers could determine their rate of loss. Four beans per square foot equaled a loss of about one bushel per acre, while just twenty beans per square foot indicated a loss of five bushels per acre. Using the new machine to minimize losses meant maximizing returns on a profitable crop.[34]

From Combines to Nostalgia

Industry observers noticed the shift from small grains to soybeans as well as farmers' abandoning their old binding and threshing tools and embracing the combine. In the fall of 1950 John Deere & Company placed an advertisement in *Farm Implement News* titled "Time Out, Old-Timers." The self-serving text evoked nostalgia and images of modernity:

This fall, throughout the nation, the old crews have been getting together again . . . this time to thresh the past, to harvest the rich yield of memory, to claim and store forever the golden grain that only fellowship can sow and only time can nourish. And we of John Deere are proud to join them in their retrospect, proud to share with them their golden memories, proud that we and our John Deere dealers have been a part of the great progress in harvesting equipment they have witnessed.[35]

The advertisement emphasized the new prominence of farmers who purchased combines. It relegated threshing to the past and highlighted festivals such as the First Annual Midwest Old Settlers-Threshers Reunion held at Mount Pleasant, Iowa, that September. Rather than playing a vital role in getting the crop from the field to the granary or elevator, threshermen were now "old-timers," belonging to a golden yesterday characterized by intense farm labor. The advertiser's new world of grain harvesting was characterized by comparative ease and speed, with the harvest requiring the labor of no more than two people or even performed alone with minimal sweat.

In 1950 combine manufacturers could afford to be nostalgic. Combine sales were on the rise throughout the 1950s. In 1952 farmers reported 71,728 combines in use on 197,741 Iowa farms, that is, just over one combine for every three farms. By 1959 there were 90,027 combines on 184,866 Iowa farms, that is, one combine for every two farms. Although the numbers of these pull-type combines for small grains and beans declined after 1959, so did the number of farms, and farmers' commitment to combining did not abate. Manufacturers introduced new self-propelled models that handled corn as well as the lesser grains, supplanting the tractor-drawn models as well as corn pickers. Farmers who participated in the reduced threshing rings yielded to combines over the course of the 1950s, by either choice or attrition. By 1960 the "old-timers" could enjoy the nostalgia of their youth and fellowship of threshing at the two-day Mount Pleasant Reunion without a month's hard work of production agriculture.[36]

During the 1960s, tractor-drawn combines also became a mere memory for some Iowa farmers. With the development of a self-propelled combine designed for the corn crop as well as small grains and soybeans, it was possible to own one machine to harvest all the seed crops grown on a farm. Farmers with the new style self-propelled combines merely changed harvesting heads in order to switch from small grains or soybeans to corn. With a self-propelled combine farmers could drive directly into the field, instead of driving over part of the crop as they did with the tractor pulling a combine, which meant that more of the crop would be harvested. Manufacturers discontinued tractor-drawn models that were not suited for corn harvest in favor of the true multipurpose machine. As one industry observer noted in 1965, the trend toward harvesting shelled corn in the field meant the end of pull-type models. Self-propelled machines accounted for approximately 35 percent of all harvesting machines shipped to dealers in the United States in 1958, but by 1964 they accounted for 51 percent of all harvesting machinery shipments.[37]

For all the hyperbole about "dry shirt" harvesting in 1940, combine advertisers based their pitch on a degree of truth. Farm families looked back with pride on their tradition of demanding physical labor, but they also anticipated a future when they could, as another John Deere advertisement commanded, "Lift the Burden of Long, Drawn-Out Harvests." Farmers themselves reported their priorities for 1951 in *Wallaces Farmer*. "This will be the last year we will cut and thresh our oats," a Jackson County farmer stated that summer. A Dallas County farmer noted that he wanted to harvest the "quickest way possible—that's for me." During the 1950s, threshing with a tractor and threshing machine became obsolete. In the 1960s, a minority of farmers replaced their pull-type combines with self-propelled combines. The decision about harvesting small grains and soybeans was no longer whether to combine but, rather, what kind of combine to use. Farmworkers, both men and women, who purchased or hired combines cut costs over competing technology, adapted the machines to meet their needs, reallocated time for other work or leisure, and claimed a place in a new dual-crop regime of corn and soybeans.[38]

Eight

From Corn Picker and Crib
to Combine and Bin

The post–World War II period was a time of tremendous change in corn harvesting and storage techniques. At the beginning of the period, farmers relied on the mechanical corn picker, a machine that removed the ears of corn from the plants in one or two rows, depending on the size of the machine. By the late 1960s many farm families had replaced mechanical corn pickers (that harvested ear corn) with combines that harvested up to four rows at a time and shelled the kernels off the ear in the field. Families who stored their entire crop of ear corn in the ubiquitous drive-through corncribs remodeled those structures and added new types of buildings in order to accommodate shelled corn. They purchased drying equipment to prepare the crop for storage or sale. This bundle of corn-harvesting technology was much more expensive than any technology contemplated by farmers in the postwar period. The conversion to combines, crop dryers, and storage buildings was expensive for Iowa farmers, especially since they already had significant investments in equipment and buildings for ear-corn harvesting. This all made the transition a slow process.

While most farmers in Iowa would have readily acknowledged the advantages of mechanical corn picking over hand harvesting in the 1930s and 1940s, they were much more skeptical about the merits of the new picker-sheller combination machines or the new combines adapted to harvest the corn crop in addition to other seeds on the market in the late 1940s and 1950s. With the ear-corn harvest, farmers would spread one load at a time on top of the previously harvested corn before returning to the field, allowing air currents to move among the ears and dry the corn enough so it could be stored without spoiling until warm weather arrived in the spring. Their storage buildings were designed for ear corn. If farmers brought in only shelled corn kernels from the field, instead of the entire ear, how would they store all that shelled corn? If they could harvest their entire crop in the course of a few days rather than a week or more, how would they keep the newly harvested but still moist corn from spoiling in the bin? Furthermore, the new harvesting and storage techniques were much more costly than the practices they were replacing.

Many farmers continued to use mechanical corn pickers for all or part of their crop throughout the 1960s, but the value of the new technology was apparent to those who wanted to maximize yields. Farmers who changed from hand picking to mechanical corn pickers as recently as the 1940s then turned to combine harvesting only in the 1960s. They invested so much in chemical and other mechanical technologies that they could no longer afford to avoid the more efficient (and costly) combine. The combine, drying equipment, and new storage structures were catalysts for other changes in the countryside, including farmers' increasing indebtedness and the expansion of farm operations.

Ear-corn Harvesting and the Mechanical Corn Picker

In the 1930s and 1940s Iowa farmers were eager to make the transition from hand harvesting to mechanical corn pickers. Harvesting ear corn by hand was among the most tedious and time-consuming jobs on the farm, according to both historians and contemporary observers. Farmers spent weeks, sometimes months, conducting their harvest. Carl Hamilton grew up during the interwar years and claimed that picking corn by hand was the worst drudgery on the farm. Each ear had to be removed from the stalk and tossed into a wagon. Hand picking was not only a long day of labor; it continued over weeks or months until all of the ears had been gathered from the field. The longer the corn stayed in the field, the more likely it was that some ears would fall off the stalks or the stalks would break in a wind, forcing the pickers to bend over to recover the crop from the ground.[1]

Many farm women were especially thankful for mechanical corn pickers. Farm families hosted itinerant corn pickers during the harvest season to help get the crop into the crib in a timely manner. Unlike regular hired men who were paid a regular wage for an entire year, corn pickers were paid by the bushel, and the family furnished room and board for the duration of the harvest. Regular hired men did other farmwork when it was raining or too muddy to get into the field. The corn pickers did not. Instead they waited indoors for conditions to improve in the field. Imogene Hamilton dreaded picking season in the age of hand picking for she would have to deal with "a bunch of corn pickers loafing around the house." Carl Hamilton noted that many of these men performed their work admirably, but "they were not always the kind of fellows you would invite in for Sunday dinner. The extra work of hosting these men as 'house guests' for a few days of nasty weather was not appreciated." Even under the best conditions, feeding pickers and cleaning up after them were onerous additions to the farm woman's regular duties. As one Wisconsin woman stated, no one mourned the end of the hand corn-picking era, "And mother mourned it least of all." Two southwestern Iowa women contrasted the new days with the old days of feeding up to seven men three meals a day for weeks on end. Instead of having meals ready before daylight and scrambling to get a pie baked by late morning, one woman whose family owned a mechanical picker went to town in the morning, helped a neighbor butcher chickens, and fed only one extra man at dinnertime.[2]

Mechanical corn pickers were prized commodities, although they were not widely available in the 1930s and 1940s. The combination of poor commodity prices during the 1930s and government restrictions on farm machinery purchases during World War II kept them out of reach for many farmers who hoped to leave hand picking behind. Those who farmed large enough acreages to justify the purchase enjoyed the benefits of rapid harvesting. In 1939 a Mahaska County farmer who harvested corn for himself and three other farmers stated, "with the machine we were at it only two weeks and finished in time to go pheasant shooting." The cover illustration of the January 7, 1943, issue of *Farm Implement News* indicated the extent to which corn pickers were treasured during the wartime labor shortage. The cartoon featured a corn picker on a flatbed railroad car arriving at a small town station under an armed guard, with the caption "The arrival of the corn picker." Led by a local law enforcement official, a crowd of farmers armed with firearms, axes, rocks, and pitchforks greeted the machine.[3]

The mechanical corn picker was far from perfect, however. Next to the tractor, it was the most complicated piece of machinery on the farm, and there was a lot of work involved in setting it up and operating it both properly and safely. Ray Gribben of Dallas County documented in his diary the amount of work required to prepare the corn picker for operations, as well as the frequency of its problems and his accompanying exasperation. In 1947 he spent half a day just searching for his operator's manual, plus a full day to get the machine in operating order. He began mounting the picker on the tractor on November 3, borrowed a chain hoist on November 4 to get the large sections mounted, and finished mounting the machine on November 6. When the machine was in position on the tractor, Gribben found that some of the husking rolls (that removed husks from the harvested ears) did not work. In Gribben's words, these parts would "take a long time to adjust." Getting a corn picker ready for the field was a process that could take hours, or even days; it was much more complicated and time-consuming than the old work of mounting the bangboard on the wagon and hitching the team of horses.[4]

Farmers were not free from machinery troubles when the harvest began. Frequent breakdowns and repairs plagued farmers who used corn pickers. Gribben spent ten days picking corn on his farm in 1947, but there were breakdowns on four of those days, which caused him to spend part or most of the day engaged in repairs. After he broke the hitch on the corn picker on November 14, Gribben drove to nearby Perry and Dallas Center in search of a replacement part. He purchased a part for a more recent model of picker in hopes that it would work but on the following day had the old part repaired in Perry. In this case, Gribben lost almost two days of work for one repair. That December Gribben assisted with repairs when a neighbor used the machine. He spent the afternoon of one day and part of another day in repairs before the neighbor broke the machine on December 13. After a flurry of repairs the next autumn, Gribben remarked, "So—the old machine is up to its old ways—a few hours work— then a day's lay off." Repeated corn picker breakdowns, both major and minor, plagued almost every farmer to varying degrees.[5]

New models were also subject to breakdowns, although they were less likely to give trouble than the older machines. Both new and old corn pickers were so complex that there were many different things that could go wrong. Joseph Ludwig of Winneshiek County incurred an average of $15.16 in expenses each year he operated his picker from 1945 to 1965, but the years immediately before the purchase of a new picker were some of the most costly in the working life of the machine. In 1945 Ludwig made eleven repairs on his picker in January, June, October, and November, which cost $24.13, and ten of those repairs were during the corn harvest. Ludwig purchased new corn pickers in 1946 and 1955 but still incurred minor repair expenses in each of the years following those purchases. Farmers such as Ludwig expected several years of low-cost operation from new machines, but inevitably repair costs mounted as the machine aged. The first major expenses for repair of the 1946 machine were in 1950, and significant repairs on the 1955 machine began in 1961 and continued to 1965, when Ludwig spent $110.83 on nine repairs in October, four in November, and a final repair of the year in December.[6]

Storage for a "Flood" of Ear Corn

As farmers grappled with the annual headaches of the corn harvest, they confronted a new and enviable problem, the increase in corn yields. The combination of using hybrid seed corn and applying larger amounts of commercial fertilizer resulted in higher yields per acre. The trend began in the late 1930s and continued during the war years. Corn yields did not increase dramatically each year, but with a slow and steady rate they grew higher in the 1940s and early 1950s than they had been for most of Iowa's history. Farmers of the 1920s expected yields from the upper thirties to the low forties, while yields in the 1930s varied widely from lows of twenty-eight and thirty bushels per acre during the drought years of 1934 and 1936 to highs in the low fifties in 1939 and 1940. With favorable weather conditions, farmers confronted improved yields and a storage problem that their parents in their wildest dreams would never have anticipated.[7]

In November 1948 the editors of *Wallaces Farmer* proclaimed, "Corn Floods Iowa Farms." It was the largest corn harvest in the history of the state, and farmers were finding they lacked adequate storage. Corn acreage was up from the 1947 crop year, but the biggest reason for the gain was excellent rainfall and heat at the proper times in the growing season and a dry fall that was ideal for the maturing crop. Corn yields had been low in 1947, with a statewide average of thirty and one-half bushels per acre, but in 1948 the state average was sixty-one bushels per acre. According to farm journalists, the permanent corncribs were filled early, leaving farmers to fill the driveways of their cribs and to construct a variety of temporary structures in order to hold the harvest. Another record crop in 1952 had farmers scrambling again for storage space.[8]

The shortage of corn-storage space was the result not just of increased production but also of a lapse in construction during the agricultural depression of the 1920s and 1930s. There had been little new construction on Iowa farms during the worst years of the Great Depression, compared to the period from 1900 to 1920 when farmers enjoyed an unfamiliar degree of prosperity. During the depression, crop prices were at record lows, and in many cases the yields were also at a record low. Many families were struggling to keep their farms, and improvements to buildings and fences were out of the question for most farmers. Aging corncribs and granaries and too few of them were the rule when prices finally recovered and production increased during and after World War II.

When the war ended, it became possible again for farmers to consider building new cribs, in large part because of federal farm policy. In 1942 Congress guaranteed supported prices for many commodities and passed the Steagall Amendment, which provided for the continuation of price supports at wartime levels for two years after the war. This promise gave farmers a degree of security they had never experienced before. Congress extended the provisions of the Steagall Amendment well into the 1950s, thus guaranteeing a degree of predictability in income that allowed farmers to invest in their farmsteads.

In the 1940s and 1950s many Iowa farmers were willing to purchase new corn-storage buildings, and the corncrib-and-granary combination was the ideal. The center driveway ran from gable end to gable end, with rooms on either side of the drive for storage of ear corn and shelled corn. The new cribs, however, differed in two major respects. First, the floor plan of new cribs was often larger than the cribs constructed during the "golden age" of farming from 1900 through the 1910s. Modern cribs also included more overhead grain storage, which meant they were often taller than the older cribs. As a writer for *Wallaces Farmer* noted in 1948, there were few "modern" corncribs in the state. The center driveways of the older corncribs were designed for horses and wagons, not trucks. With a few modifications to accommodate tractors and trucks, the old-style crib could be modernized.[9]

The cost of these new corncribs varied, but they were the most expensive kind of storage farmers could buy. As Earl Van Donselaar of Mahaska County commented, "The kind of crib most of us want costs more than a dollar per bushel of crib space if you hire the labor," a concern that prevented many farmers from building permanent cribs. Van Donselaar built a traditional-style crib but saved money by doing the labor himself, while other farmers used rough-cut local oak lumber or recycled lumber from other buildings to build their ideal crib. *Wallaces Farmer* staff members documented the wave of corncrib construction in the late 1940s and early 1950s. Many of these buildings featured curved gothic roofs made with laminated rafters or gambrel-style rooflines. As one writer noted in the early 1950s, "you see quite a number of wide-driveway, overhead bin type cribs under construction." Expenses associated with these new buildings are indicated in farm record books. In 1949 Melvin Laughlin of Hardin County built a new crib for $1,702.25, while Rudolf Schipull constructed one that same year for $6,792.25. Many farmers spent such large sums because these buildings reflected older ideas and traditions about what constituted proper corn storage. After years of privation and struggle, farmers had high yields and good prices and could construct the "dream crib of the 20s and 30s," which signified their success and perseverance. New cribs equaled or exceeded the size of the barns on many farms.[10]

Farmers who were open to new ideas about what constituted a good permanent crib could try styles designed by agricultural experts from the private sector or the land-grant colleges. The Quonset building associated with the military during World War II became a feature on many Iowa farms and was used for a variety of purposes, including storage for equipment and grain. In 1952 W. R. Mitchell of Grundy County constructed a Quonset as a granary, complete with a forced-air drying system. Agricultural engineers at Iowa State also drafted building plans for the Midwest Plan Service, a consortium of engineers from land-grant colleges that offered low-cost plans suited to midwestern conditions. The Midwest Plan Service first offered corncrib plan number 73281 in 1953. The crib was rectangular, with two pens running the length of the building that were separated by an A-frame center alley for air circulation. Hatches on the roof allowed farmers to use portable elevators to move ear corn into the crib, just as they would a more traditional crib. These buildings were suited for natural-air drying, or farmers could use a fan in one of the gable-end A openings to force air through the cribbed corn to speed drying. These cribs did not become popular, however. The timing of the new buildings coincided with the rise in harvesting shelled corn.[11]

Despite the attention focused on new structures, the most common new type of corn storage was the temporary crib, made from inexpensive materials, erected in the simplest of manners, and expected to last from one to four seasons. Farmers often used round posts or dimension lumber and snow fence or woven wire fence to contain ear corn and to keep most of it off the ground. Sometimes temporary cribs were the height of several rows of snow fence and were square, rectangular, or round, varying in size. William Beardsley of Decatur County made a floor for his snow-fence crib and even added a roof from an old steel grain bin. Other farmers such as Fred Beier of Buchanan County built a temporary crib in the driveway of his permanent crib. He allowed space on either side of the new crib for ventilation and gained the benefit of a roof that was already paid for. Melvin Laughlin prepared for his extra storage demands in September 1946 by purchasing fifty feet of snow fence or wire cribbing and lumber, which cost him $17.76. Laughlin's low investment suggests that temporary cribs were an attractive solution because they were inexpensive compared with the cribs he and Rudolf Schipull built in subsequent years. The low first cost was a way to ease cash-flow problems, although corn stored in temporary cribs was more susceptible to moisture or rat damage than corn stored in permanent cribs.[12]

Some farmers constructed portable cribs to meet their livestock-feeding needs. In the fall of 1951 Carl Anderson of Washington County built cribs ten feet wide and thirty-six feet long in a pasture in order to facilitate feeding hogs. He set posts four feet apart and braced them across the narrow width of the crib to keep it from bowing under pressure from the ear corn, then used snow fence for the sides. In 1952 he rolled up the snow fence, removed the braces, pulled the posts out of the ground, and moved the crib to a new pasture. Anderson's crib required reassembly on the new site, but other farmers around the state solved that problem. They built cribs on skids or runners, with wire mesh or lumber sides, making it easy to move empty cribs around the farm with a tractor.[13]

For all the flurry of construction of new and temporary cribs during the 1940s and 1950s, few farmers considered the implications of the new harvesting machines on the market in those years. The problem, according to one writer for *Wallaces Farmer* in 1954, was that the old-style crib was not necessarily the best thing for the future. He asked farmers if the old-style crib "will fit your corn harvesting methods and corn storage needs in the

1960s and 70s?" By the mid-1950s, farmers had more choices in harvest technology. The picker-sheller was simply a corn picker with a shelling unit attached, but the combine for corn was something new. John Deere introduced the first self-propelled combine adapted for corn in 1955. The combine with corn-harvesting attachment (called the head) enabled farmers to harvest only the corn kernels and leave the cobs in the field.[14]

Shelled-corn Harvest and the Picker-sheller

The picker-sheller was the first of the new harvesting machines to be introduced. The Massey-Harris Company pitched the picker-sheller to veterans returning from World War II in a 1946 film titled *Into Tomorrow*. The protagonist was a veteran, returning to the farm, who asked what kind of future he could expect in agriculture. The veteran was looking at the self-propelled picker-sheller, a machine resembling a corn picker in the front, but which also had a shelling apparatus on the back, to separate the cobs and the husks from the kernels of corn. Like manufacturers in many sectors of the immediate postwar economy, Massey-Harris presented an optimistic view of technology, in which people who used new products also enhanced their lifestyles. In their vision the new equipment in corn harvesting, haymaking, and small-grain harvesting offered young people a promising future on the farm.[15]

Farmers who used picker-shellers could expect several advantages over those who used mechanical corn pickers. One of the most important changes was the timing of the harvest. Farmers with picker-shellers could harvest earlier in the season when the corn in the field had a higher moisture content. Harvesting high-moisture corn allowed them to harvest more of the crop. As corn dried in the field, the ears were more likely to drop to the ground beyond the reach of mechanical pickers. The problem of ear drop became a much bigger issue in the 1940s and 1950s as farmers contended with infestations of the European corn borer, which weakened both the stalk itself and the ear shanks holding the ear to the stalk. The longer the corn stayed in the field, the greater the risk that high winds could knock the stalks over or force ears to drop, especially in infested fields. Harvesting machines also created their own losses. Pickers inadvertently shelled some of the kernels off the ear as it moved through the machine and into the wagon that trailed the picker. This problem was worse in dry conditions. Harvesting corn early at high moisture content with a picker-sheller allowed farmers to get more of the crop from the field to the crib.

Purchasing a picker-sheller made financial sense for farmers who wanted to get the most out of their harvest or who harvested many acres of corn. Harold Folkerts of Butler County invested both in a picker-sheller and in drying facilities because his corn yields were so high he would otherwise have had to build new expensive corncribs. Folkerts reasoned that the investment in a new machine would actually save money, since the cost of the picker-sheller and dryer was less than the cost of cribs. Iowa State engineers reached the same conclusion the following year when they recommended that farmers who harvested more than 125 acres of corn could use a picker-sheller for less expense than mechanical picking and storing ear corn. Howard Sparks of Story County was in an ideal situation for converting to a picker-sheller in the early 1950s. He needed to replace a worn-out picker, and his landlord invested in storage bins for shelled corn. Sparks claimed that the harvest was easier and allowed him to reduce harvest losses.[16]

The picker-sheller was especially popular in the north-central part of the state because farm families there raised corn for sale more as grain than for livestock feeding. Farmers testified about the advantages of the picker-sheller, especially its labor savings and their ability to harvest early with it. An optimistic young Webster County farmer asserted that "The picker-sheller is the coming thing. . . . It's lots less work and handling [of grain]." Albert Boes of Carroll County used a picker-sheller on 85 acres of corn in 1954 and harvested by himself. He was harvesting at 30 percent moisture content, which allowed him to get more ears while they were still on the stalks and to avoid losses due to harvesting. John Johnson of Buena Vista County also favored harvesting at approximately 30 percent moisture with his picker-sheller. In 1954 he harvested 75 of a total of 800 acres by himself—a remarkable feat compared to the hand harvesting still popular at the beginning of World War II. Similarly, Clifford and Wayne Rabe of Sac County harvested with a picker-sheller because, according to Clifford, they "don't like to work." This tongue-in-cheek reference to labor savings reflected a real benefit of the picker-sheller. Leaving the cobs in the field meant there was less shoveling at the crib, and hauling costs were reduced by half.[17]

Moving shelled corn rather than ear corn meant there was much less bulk to haul and fewer trips from the field to the storage building, a major advantage during a period of scarce labor. It was easier to utilize family labor or part-time help rather than rely on a hired man to do the hauling. Fewer trips also meant less fuel consumption. Iowa State extension specialists calculated that it cost farmers $0.25 per bushel to transport shelled corn compared to $0.32 for ear corn. This savings might seem small, but it becomes significant when multiplied by thousands of bushels. A farmer who harvested 60 acres of corn at fifty bushels to the acre in 1962 might believe that $210.00 of savings was enough to warrant investing in new corn-harvesting technology, but a farmer with 200 acres of corn at fifty bushels to the acre could save approximately $700, roughly 10 percent of the purchase price of a new combine suited for corn harvesting.[18]

Reducing labor requirements and hauling costs were only some of the advantages of new technology of shelled-corn harvesting. An equally important issue was the need to get as much of the crop into storage as possible. Concern about getting every kernel became even more compelling in the 1950s because it cost more to raise a crop than it did before World War II. Farmers applied commercial fertilizer to crops in order to obtain higher yields than from the traditional crop-rotation systems. Pesticides also enhanced yields, because they controlled the weeds that competed with crop plants for moisture, sunlight, and soil nutrients and insects that preyed on the root systems and stalks of corn and other crops. These expenses made it all the more important for farmers to preserve as much of the crop as possible.

The development of the combine to harvest corn in addition to small grains and soybeans was another tool to minimize harvest losses and to harvest earlier. In 1950 agricultural engineers worked to develop the corn head to mount on a self-propelled combine. In 1955 John Deere Company introduced the first corn head on the market, the Number 10. The new corn head could be mounted on self-propelled combines in the place of the attachment head for cutting small grains or other seed crops such as soybeans. "Corn combining is here—and here to stay," boasted John Deere advertisers in the summer of 1956. According to the company, farmers who used combines cut their shelled-corn loss by 75 percent and ear-corn loss by 50 percent compared to mechanical pickers.[19]

The reality of combining may have been less awe-inspiring than advertisers claimed, but it was still impressive. Farmers who used pickers expected to lose approximately 10 percent of the crop in the field, which meant they would either have to go back and pick up ears off the ground by hand or have to turn livestock into the field to consume what was left behind. Early harvesting with a combine at higher moisture content (25–30%) provided an important advantage over using a picker to harvest at approximately 20 percent moisture. With a picker-sheller or combine, farmers could expect to lose only 6 percent of the total crop. Clarence Wolken of Marshall County used a combine in the late 1950s and claimed that "You can start harvest earlier in the fall, and get away from ear drop." In 1963 Paul and Gordon Christensen of Boone County purchased a new corn combine. They had used a picker-sheller since 1959 and, according to Paul Christensen, "saved a lot of corn by harvesting earlier." He also asserted that "we even saved more corn with the combine." Marvin Bacen of Humboldt County testified that "You can pay for a combine just from the difference in field losses." By spending more for a new harvesting machine, farmers could save a greater percentage of their crop and earn more money.[20]

Harvesting shelled corn also cut costs after the harvest, too, since it was no longer necessary to hire custom shellers to visit their farms periodically and shell the ear corn. Shelling and grinding corn was the best way to prepare it for feeding, since livestock utilized more of the nutritional value of the grain when it was cracked. Shelling was not normally considered a direct cost of harvesting, since farm families incurred shelling costs over the course of the year as they depleted feed stocks. It was, however, an avoidable indirect cost that was no longer necessary if farmers harvested shelled corn.

For all the advantages and promotion of picker-shellers and combines, farmers did not rush to adopt this new technology in the second half of the 1950s and the early 1960s. Most farmers simply did not have the number of acres to justify purchasing a combine unless they planned to do extensive custom harvesting. Two Grundy County farmers interviewed at the Farm Progress Show in 1959 estimated that a farmer would "need about 300 acres [of corn] to justify owning one." That same year agricultural engineers estimated an acreage threshold for combine ownership of 200 acres, which suited the needs of only a small minority of Iowa farmers. The discrepancy between the perception of the two men at the Farm Progress Show and the carefully calculated figure of extension experts reveals the gulf between the majority of farmers who could not envision a profitable use for a combine and the minority who could afford to study the acreage threshold. As expected, farmers with large acreages were often the first to purchase combines. Clarence and Lester Wolken of Marshall County began using a combine in the late 1950s to harvest 150 acres of corn. A survey conducted in 1965 indicated that approximately one-third of farmers like the Wolkens who planted 150 acres or more of corn would harvest at least part of their crop with a hired or purchased combine. By contrast, only 10 percent of farmers with less than 150 acres of corn that year planned to harvest shelled corn. Most farmers agreed with the Grundy County farmers at the Farm Progress Show—that combine harvesting was impressive but was also someone else's business.[21]

Combine harvesting began in earnest during the mid-1960s, almost ten years after combines for corn first came on the market. In 1964 Iowa farmers harvested only 13 percent of their corn acres with combines, but in 1967 they used combines on 32 percent of the state's corn acreage. In 1968 farmers used combines for 35 percent of Iowa's 9.7 million acres of corn, with picker-shellers responsible for harvesting another 8 percent. By 1972 farmers were harvesting 63 percent of the corn crop with combines and another 7 percent with picker-shellers, a significant change from the mid-1960s.[22]

High costs accounted for part of farmers' initial reluctance to purchase combines. Joseph Ludwig's farm records indicate the high capital requirements of making the transition from mechanical corn picker to combine. Ludwig purchased two new corn pickers in the years after World War II, the first in 1946 and the second in 1955. These machines cost $943.23 and $1,300.00, respectively, making them some of the most expensive machines Ludwig ever purchased for his farm. In 1966, however, he invested $8,600.00 in his first self-propelled combine for harvesting corn, small grains, and soybeans. While this machine was capable of harvesting a greater variety of crops than his older corn pickers, the expense was more than the cost of a new corn picker and new small-grain combine put together at 1966 prices. Ludwig's new 1966 machine had more capacity than either a corn picker or a tractor-drawn combine, but it was still a sizable investment when compared to the older technology.[23]

Just as the first cost of a combine was greater than that of the competing technology, so were the operating costs. The impact of the new high-priced machines on farm operations is clear from an examination of maintenance and repair costs. Corn picker repairs were often nickel-and-dime expenses. Farm records show that many farmers often spent less than a dollar or two at a time, reaching a total of only a few dollars per year. Joseph Ludwig spent an average of just $15.00 per year on corn-picker repair from 1945 to 1965. Combine repair and maintenance, by contrast, were costly. Ludwig spent an annual average of $98.76 from 1966 to 1970 on his new combine. Iowa State economists estimated that repairs and depreciation costs totaled approximately 14 percent of the original cost of the machine. Calculated this way, Ludwig's $8,600 combine was actually a $9,804 investment, excluding interest.[24]

Storing Shelled Corn in Dryers, Remodeled Cribs, and Bins

The most significant obstacle to harvesting shelled corn with combines or picker-shellers was the problem of how to store the shelled corn. The majority of farmers were equipped to store ear corn, not shelled corn. The writer for *Wallaces Farmer* who asked farmers in 1954 if the old-style crib "will fit your corn harvesting methods and corn storage needs in the 1960s and 70s?" addressed an important issue. While the capital outlay for new harvesting equipment was high, it was compounded by the fact that farmers who harvested shelled corn needed expensive new storage buildings and special drying equipment. In the early 1960s new drive-through-style cribs cost as much as $1.25 per bushel, and metal bins cost approximately $0.35 per bushel. The Gerlach family of Story County built an old-style crib in 1956 but converted it to shelled-corn storage just three years later. As Ralph Gerlach explained, "We had no idea when we built it to hold ear corn that we would switch to using a picker-sheller so soon." Most farmers were trapped with buildings designed for an earlier era. According to Ken Smalley of Johnson County, "Farmers already had these cribs and they felt like they had to use them." So farmers continued to use the harvesting machines that matched their infrastructure.[25]

Remodeling old corncribs was one of the most attractive ways to obtain storage for shelled corn throughout the 1950s and 1960s.The common drive-through cribs were the most obvious choice for conversion to shelled-corn storage. If the crib was in good condition, farmers and experts reasoned, it was simply a matter of strengthening the floors and walls while making it tight enough to hold loose grain. To convert a crib, farmers first added lumber wind-bracing across each half of the crib. Lining the crib with plywood also helped brace the structure against wind shear. To keep the building from expanding outward under the pressure of the shelled corn, they added tie-rods through the crib secured to lumber on the outside of the crib. Finally they enclosed the building with plywood, boards, or sheet metal. If the crib was not on a concrete foundation, the floor joists needed to be reinforced to carry the heavier load.[26]

Farmers who wanted to change storage facilities liked the low cost of crib remodeling. Harry Wassenaar of Jasper County explained, "Converting our old crib was the cheapest way" to get more storage and convert to harvesting shelled corn. Delmar Van Horn of Greene County was one of the first of many farmers featured in *Wallaces Farmer* who remodeled cribs. Van Horn claimed that he spent $470 in materials to convert his entire four-thousand-bushel ear-corn crib to a building that would store eight thousand bushels of shelled corn. The total cost for new construction, including labor, was approximately twenty-five cents per bushel. Clarence Wolken and his son Lester of Marshall County converted a crib in 1958 for approximately seven cents per bushel, by using salvaged lumber and their own labor. In 1959 they hoped to remodel another crib for five cents per bushel for materials. Without salvaged lumber, farmers could expect to spend approximately fifteen cents per bushel for supplies. As long as the labor cost did not exceed ten cents per bushel, remodeling was cheaper than new construction. These kinds of savings made the transition to shelled-corn storage affordable for many farmers.[27]

Remodeling corncribs was only one option for farmers who wanted to cut costs on construction of shelled-corn storage. Farmers could convert obsolete buildings into storage space. In 1954 Floyd Doxtad of Ida County invested one hundred dollars in reinforcing an old hog barn for shelled-corn storage. The iconic barns of the horse-powered mixed-farming era were also good candidates for remodeling. Willis Scott of Hamilton County remodeled his barn, designed for hay storage, to grain storage. He removed a dozen horse stalls and feeding areas for cattle to make room for forty-two thousand bushels of shelled corn. He removed the second-floor haymow, poured fifty yards of concrete to make a four-inch-thick floor to support the grain, and divided the building into three grain-storage bins. Danny Bohrofen of Keokuk County owned an unused barn with an overhead haymow, and horse stalls and milking stanchions on the ground level. "The barn was idle," he noted, "and I was paying taxes on it," making it a candidate for either demolition or remodeling. Bohrofen altered the barn in stages, first converting the stalls and stanchion areas into bins for sixteen thousand bushels, then adding overhead bins in the mow to hold seventy-five hundred bushels. At harvest time he used augers to fill the bins. These kinds of modifications allowed farmers to make the transition from ear-corn storage to grain storage in an economical manner.[28]

Modification of older cribs and farm buildings minimized farmers' expenses, but new grain-storage buildings such as grain bins also became common on farms as more farmers harvested shelled corn. Steel grain bins were distinctive buildings, characterized by their cylindrical shape, conical roof, and corrugated steel siding. Bins adapted for drying often had perforated floors to allow heated air to be forced up through the grain. Unlike corncribs, there was never any intention of holding ear corn in these new-style bins or segregated spaces for ear, shelled, or cracked corn. They were strictly for shelled corn, grain, or soybeans. Although grain bins were on the market in the early 1900s, few farmers had invested in them at that time because most corncribs or granaries included enough capacity for as much grain as farmers produced in the era of mixed farming and corn picking by hand.

Grain bins were structures for industrial agriculture, just as combines were machines for industrial farming. The largest and most highly mechanized farms were the most likely sites for the new grain bins. A 1956 advertisement for the Behlen Manufacturing Company for cribs, bins, and dryers indicated the size of operation in which grain bins could be most successfully utilized. The advertisement featured Cedar County farmer Carl Levsen and his three adult sons who farmed 700 acres, of which they planted 250 acres in corn. They installed eight grain bins with a total capacity of twenty-five thousand bushels and owned seven more thirty-two-hundred-bushel bins that they had not yet assembled. According to the advertisement, Levsen emphasized cutting operating costs and marketing quality products. "We call it industrialization," Levsen and the advertisement copywriters proclaimed, suggesting that contemporary farm operators would have to borrow from the world of business management to survive. The Behlen Company emphasized the role that grain storage could play in industrializing agriculture, and in changing the landscape of the farmstead.[29]

While the Levsen family represented the "think ahead" mentality, most farmers resisted the move to new storage. A 1958 poll of Iowa farmers indicated that 98 percent stored ear corn in cribs while only 21 percent stored shelled corn in bins. The numbers total more than one hundred because some farmers stored corn both ways. When asked if they were considering any changes in their storage systems, 83 percent of Iowa farmers responded in the negative. Their lack of interest in change reflected the fact that most farmers did not have the acreage to justify purchasing a picker-sheller or a combine and therefore lacked incentive to change. Even farm-management specialists giving advice in a *Wallaces Farmer* column counseled farmers with average or small-sized farms to avoid the shelled-corn harvest and grain bins. "Don't invest too much in buildings and equipment on a 120 [acre farm]," counseled a bank president from

Schleswig, Iowa. Ear-corn harvesting and temporary cribs made from snow fence were a better option for the majority of Iowa farmers in the 1950s and early 1960s.[30]

Even farmers who did use bins continued to harvest ear corn, as the survey results from 1958 indicate. This middle way, using some new storage while continuing to use the old also, was the only way that made sense to farmers who already possessed structures for ear corn but wanted to realize the advantages of earlier harvesting. In the face of mounting corn harvests with increasing yields, constructing some new storage was a hedge against wet conditions at harvest time. Arnold and Francis Krueger of Hardin County owned conventional corncribs and one twenty-one-foot-diameter metal bin that held forty-four hundred bushels. Francis Krueger liked the idea of having storage for that much corn. "If all the rest of the corn is wet," he stated, "we know we can count on the corn in the new bin keeping." They could also harvest earlier and dry the grain in the bin, then move the dried corn into the overhead bins of the corncrib, thus saving a greater portion of the crop.[31]

Farmers erected more grain bins throughout the late 1950s and 1960s. The 1958 poll indicated that a substantial majority of farmers were not considering any changes to their corn-storage systems, but 12 percent of respondents desired change. Of those who envisioned a new system, 43 percent wanted to install bins. The low cost of metal bins compared to other types of storage made this the most attractive method. Farmers modified older buildings and cribs to ease their storage needs, but in fact the yearly corn crop was increasing in size faster than farmers could modify their old bins. An older crib that held four thousand bushels of ear corn was fine for a farmer as long as yields were around fifty bushels per acre (approximately the average yield in Iowa for the period from 1941 to 1949). Farmers could hold their entire crop from 80 acres in a conventional crib. But by the late 1950s, yields varied from sixty-two to sixty-six bushels per acre, and in the early 1960s yields ranged from seventy-five to eighty bushels per acre. Eighty acres of corn yielded as much as sixty-four hundred bushels per acre, leaving the farmer twenty-four bushels short of storage space. Farmers who wanted new storage space needed bins to take advantage of combine harvesting, although harvesting moist corn and then keeping the corn dry became more important issues as farmers began to harvest earlier and move away from harvesting ear corn.[32]

The development that allowed farmers to harvest and store shelled corn was the crop dryer. Farmers who harvested corn at 25–30 percent moisture content needed to reduce the moisture to 13 percent in order to prevent grain from rotting in storage. Drying corn had not been such a problem when farmers harvested corn by hand over the course of weeks, even months. In the 1940s and early 1950s, farmers' desire to get corn out of fields infested by the European corn borer before ears dropped meant they harvested wet corn. Farmers who tried to store wet ear corn experimented with dryers to solve these problems. As soon as farmers used picker-shellers and combines to speed the harvest, drying corn became a necessity. Ray Hayes of Crawford County summed up this viewpoint in 1958. "I picked [ear] corn early last year and dried it. I liked the results," he stated. "Now I'm looking forward to a picker-sheller."[33]

There were few commercial drying units available in the 1940s to help farmers such as Ray Hayes. Dry conditions prevailed during the 1930s so there was little need to dry grain. The few farmers who wanted to dry corn improvised by using a tractor to power a threshing-machine fan to blow air through a canvas duct into the crib. Those who wanted heated air for drying enclosed the tractor engine with canvas to use the engine's heat to warm the forced air. These temporary expedients helped some farmers dry ear corn, while other farmers installed wire or wooden ventilators inside their cribs to allow for greater air circulation through the crop. Ventilators, however, reduced the amount of storage space available for corn and forced farmers to store more of their crop in temporary cribs.[34]

Farm journalists first discussed crop dryers in the late 1940s as farmers dealt with extremely moist ear corn at harvest time. In 1945 farmers experimented with placing ventilators in crops as well as forcing air through the stored corn. Farmers who used forced air used an oil burner and fan to push air into ventilators in cribs. On one Cherokee County farm, a dryer helped bring a load of 42 percent moisture corn down to 33 percent in two hours, which was promising news for farmers who risked losing a crop to mold. As long as farmers could get the corn to approximately 20 percent moisture by the time the crop froze, it would last until the spring thaw without spoiling. Most farmers, however, were not so fortunate. In the spring of 1946 much of that stored corn was still so wet that a writer for *Wallaces Farmer* observed, "A lot of black and moldy corn is going to be offered to hogs" in forty counties in northern and central Iowa that year. An Agricultural Adjustment Administration survey of one thousand corncribs in this region indicated that almost half the corn was higher than 20 percent moisture, which meant it was likely to spoil in the coming warmer weather.[35]

Some farmers experienced success with drying ear corn in the late 1940s. In 1949 a serious infestation of European corn borers was followed by a windstorm in October that snapped the brittle ear shanks of the corn and knocked the ears to the ground. In some cases twenty bushels of corn per acre lay on the ground, out of reach of mechanical pickers. Iowa farmers salvaged approximately 27 million bushels by gleaning fields by hand while livestock cleaned up approximately 45 million bushels. Farmers with corn in the field suffered, but those who had harvested early at high moisture content and then dried their corn avoided some of the worst damage. Guy Coulter, a Grundy County landlord, picked and dried his crop before the windstorm of 1949, but his tenant's portion of the crop was still in the field during the storm, which cut the yield on the standing corn by a total of 350 bushels.[36]

The 1949 windstorm made a big impression on farmers. Denton Myers of Humboldt County considered purchasing a drying unit after the storm. In the 1950 season he planned to harvest ear corn early at high moisture content and dry the corn in his cribs. As the fall progressed and the moisture content of the corn dropped, he began to field shell the crop and store shelled corn. Crop dryers helped farmers make the transition from ear-corn harvesting and storage to shelled-corn harvesting because they allowed farmers a degree of flexibility in harvesting that previously the farmers did not have.[37]

Throughout the 1950s, farmers boasted of the benefits of early picking and drying the crop with portable units called batch dryers. Just before picking season in 1954, a farmer warned that those who did not pick corn early were "likely to have trouble" with dropped ears because of severe European corn borer infestations. Walter Cramer of Wright County used a batch dryer to dry shelled corn before storing it. He harvested with a picker-sheller when his corn tested at 28 percent moisture and then dried "batches" of 335 bushels of grain at a time. Merrit Wassom of Sac County declared that his harvest losses were so low after picking early and drying that "There wasn't enough corn left in the field to keep a goose alive." The costs associated with drying shelled corn in the fall were offset in 1955 when farmers with ear corn paid to have their crop shelled. Warm weather in March and April of 1958 and the onset of mold growth in cribs and bins was powerful evidence that renting or purchasing a dryer was necessary to save stored corn.[38]

The ability to dry shelled corn at harvest time and avoid storage problems was an incentive for farmers to use picker-shellers and combines. In 1958 Ray Hayes of Crawford County claimed that after a year of drying ear corn he was "looking forward to a picker-sheller." The 1959 harvest season was wet, which meant farmers could not get into fields. Many farmers were left with corn in the fields that was too wet to harvest while those who picked early had wet corn in the cribs. "I decided early this fall to let my corn stand in the field until it was good and dry," noted a farmer from Van Buren County. "Now I'm not so sure I did the right thing." Many farmers continued to take their chances that the weather would be suitable for harvesting and storing ear corn, but a minority harvested shelled corn and invested in dryers, or rented them, in order to minimize risk to the crop.[39]

A new innovation called the continuous-flow dryer was even faster than the batch dryer. This was a portable dryer that continuously moved the grain in the dryer, drying faster than the batch dryer, which simply held the corn in a chamber. Farmers added wet grain to the top, which was dried by forced air at 180–220 degrees Fahrenheit moving through the tubes in the grain. Fans forced unheated air through the grain at the bottom of the dryer to cool it to a safe storage temperature. By moving the grain during the drying process, continuous-flow dryers allowed farmers to add or remove grain during the drying period. This type of dryer was expensive and best suited to farmers who harvested large amounts of shelled corn, since they could keep the combines running without having to wait for a batch to be removed from the dryer. Experts suggested that farmers who harvested more than thirty thousand bushels per year could best utilize this system.[40]

Some farmers equipped grain bins with dryers instead of using portable machines, which gave them several options at harvesttime. Farmers could simply use the grain bin as they would a batch dryer, loading from two to four feet of corn into the bin, running the dryer overnight, and then unloading the corn to be stored elsewhere. This was known as batch-in-bin drying. Farmers could also dry corn and store it in the same bin one layer at a time. This was called multiple-layer drying and was more time-consuming than batch-in-bin drying but required less handling. The first layer was approximately five feet deep. As soon as that layer was dry, a foot layer could be added. After three or four days, depending on moisture content, another layer could be added.[41]

To avoid the costs of drying equipment or of new storage buildings, it was possible to hire commercial dryers, also, or to store the crop at grain elevators that provided drying service. The advantage of commercial storage was that farmers could use their time for tasks other than drying, such as fall tillage or fertilizer application. The grain elevator assumed the risk and management of the drying process. The major disadvantage of commercial storage was that many elevators had limited storage space at harvesttime, which put farmers at risk of having a wet crop and no place to put it.[42]

The variety of new techniques, machines, and structures made grain storage and drying more complicated over the course of the 1960s. Just as farmers who dealt with herbicide and insecticide faced an ever more complicated array of products, combinations of products, and restrictions on the use of products, farmers who used dryers found that corn drying also required careful management. They hoped to prevent problems associated with under-drying and over-drying. Mold was the obvious problem with wet corn, but overheating or heating too quickly resulted in cracked grain, which also allowed mold to grow. Corn that was to be sold in the fall needed to be dried to only 15 percent moisture while corn to be stored for a year needed to be 13 percent. Sometimes drying was uneven, which meant that some of the corn might stay too moist and spoil.[43]

Farmers took initiative in addressing the problem of uneven moisture content in their new-style bins designed for drying. In 1962 Eugene Sukup of Franklin County bought his first grain bin to dry and store shelled corn. He was not satisfied with the results. Sukup found that pockets of grain did not dry properly and were ruined. He designed a system to break up those pockets of grain that were not drying properly using an old coal-stoker auger from a furnace mounted in an electric drill suspended from a chain at the top of the bin. The drill stirred the grain and broke up any pockets that were too moist. Sukup filed for a patent and began to sell his "Easy Stir Auger" to implement dealers and farm-equipment companies. Elmer Horstman of Hancock County also experimented with stirring devices in the mid-1960s. He used his bin as a batch dryer and stirred the corn during the drying process, cutting drying time by half and increasing the capacity from one thousand bushels per batch to twenty-four hundred. Increasing drying speed meant decreasing costs. Without the stirring unit Horstman spent six cents a bushel for electricity and fuel to dry the crop, but with the stirring unit he reduced those expenses to under four cents per bushel. Horstman's stirring unit allowed for uniform moisture content throughout the grain bin, achieving a goal shared by all farmers who used artificial drying and grain bins. Tests by Iowa State agricultural engineers confirmed what farmers already knew or suspected—that stirring was a fast and effective way to store the highest-quality grain.[44]

Farmers hoped that the increased production they gained by harvesting shelled corn early and drying it instead of harvesting ear corn late would pay for the increased costs of artificial drying. Cost figures ranged widely, depending on the size of operation and type of drying outfit they used, but for many farmers drying became a regular part of the corn harvest. Joseph Ludwig purchased a portable dryer in 1952 and used it in his crib for ear-corn drying in 1952, 1953, and 1954, spending an annual average of $121.86 for heating oil. He also used his portable dryer on a custom basis and earned $40.00 over the course of those three years. Ludwig did not dry corn again until 1958, and then only on a limited basis. In 1965 he installed a crop dryer and LP fuel tank for his grain bin and used it every year up through 1970, spending an average of $145.65 each year on fuel. Ludwig made the transition from drying corn on an intermittent basis to making it a regular part of his operation. These costs made up a big part of the total expense of raising a crop. In 1967 one Wright County farmer calculated that of the $81.26 he spent to raise 150 bushels of corn per acre, $11.25 of that cost was in drying, making it the second most expensive part of raising an acre of corn behind fertilizer. Combine harvesting accounted for 10 percent of the cost of raising an acre of corn, making it the third most expensive part of the operation.[45]

Farmers did more drying during the 1960s regardless of the method and in spite of the expense. In 1964 they dried 86 percent of the corn crop in the crib without artificial drying. Only 13 percent of farmers used artificial drying on the farm while 1 percent had their crop dried commercially off the farm. By 1970 there was a significant increase in artificial drying.

Farmers used cribs and natural-air drying for 57 percent of the corn crop and artificial drying on the farm for 41 percent of the crop with 2 percent of the crop dried off the farm. Custom drying boomed as farmers who invested in crop-drying equipment hoped to pay for their equipment more rapidly or gain extra income. Cyril Tiefenthaler of Carroll County invested eighty thousand dollars in grain-drying and -storage facilities to handle his fifty thousand bushels of corn and another hundred thousand bushels for neighbors in 1969. Tiefenthaler explained that using his equipment on a custom basis allowed him to obtain the top-quality equipment for himself and his two sons.[46]

Combines in Ascendance

Foster and Madeline Mason's farm inventory from 1970 reveals the level of investment needed to conduct this new kind of harvest. The Mason family did not own a combine, but they did invest in corn drying and storage facilities over the course of the 1950s and 1960s. In addition to two wooden granaries and three corncribs constructed in the 1950s, the Masons owned two metal bins. In the 1960s, they purchased both metal bins, one of which was a thirteen-thousand-bushel Stormor model. The Stormor bin cost $4,951 in 1965 and was second only to a new John Deere tractor in terms of value on the Mason's inventory of machinery and buildings. They owned several pieces of drying equipment, including a Butler manufactured dryer, two propane fuel tanks, and two stirring devices to aid the process of drying corn in bins. The value of these items as of January 1971 was $7,596, which comprised 17 percent of the total value of the Masons' farm inventory. When the value of the two tractors is removed, the value of the drying facilities comprised 21 percent of the tools and implements, making drying equipment the most expensive items devoted to a single process on the farm.[47]

Government programs helped farmers change their harvesting and storage practices by providing loans to farmers who stored grain as well as to those who wanted to build storage facilities. Many farmers participated in the Commodity Credit Corporation (CCC) program, which loaned farmers money to keep their crops on the farm until prices rose. The CCC, an agency of the USDA, allowed farmers the option of storing corn in bins or cribs at a low fee in exchange for government loans. A government inspector measured the storage area to determine storage capacity and then sealed the stored crop with a paper label across the door to prevent tampering. If prices advanced above the loan rate, farmers could sell their crop and keep the difference between the two prices. If prices failed to rise above the loan rate, farmers forfeited their crop to the U.S. government and kept the loan payment. Officials from the Production and Marketing

Administration (which replaced the AAA during the Truman administration) provided an incentive for farmers to reseal their corn in on-farm storage. They provided a payment of thirteen cents per bushel for resealing corn in 1953, with fifteen cents per bushel in 1955, figures that amounted to almost half the cost of constructing new grain bins. By the early 1950s, one-fourth of all Iowa farmers wanted to build or buy new corn-storage space, with another 7 percent willing to settle for temporary cribs. Those farmers who planned to build new storage space had a special incentive to construct modern storage facilities that included grain dryers.[48]

Government officials who were in charge of making CCC loans did not want farmers' crops to spoil in the crib or bin. If all or part of the crop spoiled, the farmer would not be able either to repay the loan or to forfeit a damaged crop to the government. As early as 1949 the USDA began to make low-interest loans to individuals and groups of farmers who purchased grain dryers. The loans covered up to 75 percent of the delivered cost of the dryer. Terms were generous, with three-year terms at 4 percent interest. Corncribs or bins for drying could be constructed with loans from the Agricultural Stabilization and Conservation Service, with a loan amount of up to 80 percent of the cost of construction. These programs continued into the 1960s. In 1965 construction loans for bins were available at 4 percent interest for five years, with a one-year grace period before the loan was classified as delinquent. The government even sold surplus CCC-owned grain bins to farmers. That year the federal government prepared to auction 29 million bushels of storage capacity, providing many farmers the opportunity to obtain low-cost bins. Rules for the 1967 and 1968 programs allowed individual farmers to borrow up to twenty-five thousand dollars to construct storage and drying facilities, although loans of over ten thousand dollars required a real-estate mortgage. Joseph Ludwig of Winneshiek County utilized this program to help him finance new bin construction in the late 1960s.[49]

By 1969 the increasing number and capacity of grain bins and drying systems was testimony to the growing popularity of combine harvesting and new grain-storage techniques. Bins as large as forty feet in diameter were under construction in the early 1960s, and there was no indication that the trend toward more and larger bins would be reversed. The Wiemers family operation located in Marshall County serves as an excellent comparison with the Levsen family farm (discussed earlier). The Behlen Manufacturing Company featured these families in advertisements in the pages of *Wallaces Farmer* thirteen years apart. In the 1969 advertisement, the Wiemers' operation included at least seven storage bins, five of which dated to the mid-1960s, and a batch dryer and dump pit for unloading that were added in 1967. In 1968 they purchased a continuous dryer and two large storage bins, which altogether gave them a total storage capacity of fifty-seven thousand bushels. The Levsen family, by contrast, had only twenty-five-thousand-bushel capacity for their seven-hundred-acre farm in 1956. In 1969 Orval and Gene Wiemers had over twice as much corn-storage space for their thousand-acre farm. Increasing crop yields and the expansion of operations contributed to the need for more grain storage. While the storage needs for both families in the advertisements were different, the trend toward more and larger storage units in order to accommodate the increased yields and the larger scale of farming was clear in both advertisements.[50]

Farmers who cultivated large acreages of corn used their combines profitably, while those who harvested small acreages were best suited to ear-corn harvest and storage. The combine was most profitable on farms with more than eight thousand bushels of corn plus additional acres in soybeans and possibly even small grains. Farmers such as Joseph Ludwig, who purchased a combine in 1966, used it for his own corn and soybean harvest and also performed regular custom work in the late 1960s, spreading the cost of his investment over more acres. But a combine could make sense only to farmers who were willing to invest in a grain-drying system. As the Fisher family of Hamilton County learned, replacing a two-row combine with a four-row model could make even a relatively recent drying system obsolete. They used a batch dryer and several small bins as long as they harvested with the small combine, but when they bought the four-row combine they invested in a large bin for layer drying. One of Fisher's sons explained that "the big bin gives us the extra capacity boost we need to get all our corn harvested in ample time."[51]

The prominence of combine-harvesting machines in industrial agriculture can be seen in efforts to commemorate corn harvests from the past. Just as Midwest Old Threshers' reunions in Mount Pleasant provided a forum for farmers to relive the steam- and tractor-powered threshing after it had become obsolete, farmers cooperated to remember older methods of corn harvesting and to celebrate the change. A new open-air museum called Living History Farms located on the edge of Des Moines hosted its first corn-harvest festival on October 24–25, 1970. The event featured demonstrations of hand cornhusking, one- and two-row mechanical pickers, a modern combine, and various models of hand-powered corn shellers. Farmers from central Iowa volunteered to operate the machines and discuss them with museum visitors. In addition to giving farmers the chance to reminisce about the old days, the museum assumed an educational mission "to show the younger set how it used to be done" and to "compare the best of yesterday's methods with today's mechanization." People who participated in the changes in harvesting appreciated the significance of the transition to more capital-intensive farming. They also taught a younger generation, who usually had little or no concept of the nature of the previous generation's labor-intensive work of corn harvesting.[52]

Combines, drying equipment, and structures for shelled-corn storage were much more expensive than pickers and cribs, but more farmers were using the combine on more acres by the early 1970s. The claims made for the big combines by promoters and farmers were true. Farmers could harvest earlier to minimize losses in the field, a critical consideration when they spent increasing sums of money on the chemical cocktail of fertilizer, herbicide, and insecticide. They invested their labor and the cost of fuel, machinery, and seed to get their most important crop in the ground and then helped the crop reach maturity by applying chemical fertilizer and pesticides. A growing number of farmers viewed high harvest costs as necessary to justify the other expenses of making a crop. They needed to reduce harvest losses, to harvest early, and to do so with the least amount of labor, which made increased investment a strategy to minimize risk.

Conclusion

What did farmers think about the changes they had wrought? Those who consented to be interviewed for this project recognized the revolutionary nature of their times. A retired county extension director stated that "agriculture is a different animal now. . . . It's just a different environment altogether than it was when I started." Reflecting on the tremendous scale of modern agriculture and the demise of "family" farming became a regular refrain at the conclusion of the interviews. The most common observation from the generation who lived through this transformation concerned the rapid pace and the scale of migration off the farm, a development that was both cause and effect of agricultural industrialization. The extension director who began his career in the early 1960s commented that even then "there were just a lot of farms . . . you could drive up and down country roads and see a lot of farms there," with families who worked the land. The contrast between the two eras was stark, even to contemporaries such as *Wallaces Farmer* editor Don Murphy, who claimed in 1968, "It ain't like the good ole days!"[1]

Murphy recognized the tremendous changes that had occurred in just one generation, the main change being that farmers' lives in the late 1960s were no longer characterized by such intense physical labor. People's recollections of farm life in the 1940s and early 1950s invariably included descriptions of the hard work at threshing time, the seemingly endless round of cultivating corn, and the tedium of picking each ear of corn by hand and cribbing it, one scoop-shovelful at a time. These people expressed little nostalgia about what one farmer called the "arm-strong" method of farming. They viewed the new technology through the lens of the farmwork routines, living conditions, and crop and livestock cycles that they had endured. For most of them the technological changes were generally positive.

There were concerns, however, about some of the costs of technological change. A twenty-first-century study of elderly farmers by the National Institute of Environmental Health Sciences suggested a link between Parkinson's disease and continuous exposure to chlorinated hydrocarbon insecticides. In 2005 the son of an Exira farmer recalled during his boyhood watching his father empty bags of chemicals and seeing the dust fly into his father's unprotected face. "Nobody paid any attention to that stuff then," he noted. Today's elderly farmers, according to a Des Moines neurologist, "are extremely tuned into this. . . . They're always asking, 'Could this have hurt me?'"[2]

Questions persist about other environmental costs of farming. In 2006 the *Des Moines Register* reported that, according to Iowa State University studies, Iowa's streams and rivers ranked as the most fertilizer-polluted waterways in the world. Agricultural sources accounted for only a portion of the nitrogen and phosphorous in the water, but Iowa's predominantly rural character implicated farm fertilizer as a significant contaminant. In April 2006 the Iowa Department of Natural Resources (DNR) and the U.S. Environmental Protection Agency (EPA) ended an amnesty program giving cattle feeders five years to install manure-control systems in compliance with the 1972 Clean Water Act. At the conclusion of the amnesty period, approximately 25 percent of the 150 feedlots large enough to require federal permits had not taken remedial action. Furthermore, approximately half of the 1,300 feedlots that were too small to require permits were also in violation of manure-management laws. Representatives from the livestock industry emphasized the amount of money producers invested in good faith efforts to obtain permits, construct facilities, and design manure-management plans. Critics, including some farmers, pointed to repeated fish kills caused by manure runoff. A man who had farmed from the 1930s through the 1970s chided modern farmers for using too much herbicide and suggested they reacquaint themselves with the rotary hoe. He neglected to mention that contemporary farmers were managing increasingly larger farms with less help, but his statement reflected a degree of pride in the many hours he had spent on a tractor and was a gentle rebuke of today's farmers for their reliance on chemical technology.[3]

This study highlights the extent to which the ecosystems and physical landscape of Iowa farms changed as a result of many thousands of small decisions concerning the use of technology. None of the farmers studied here ever wanted to leave pesticide, manure, or fertilizer residue in the land or to contribute to rural depopulation. Rather, the technological changes came about because they lessened the physical demands of farming and cut labor costs. Farmers' decisions were also in response to changing conditions off the farm. Federal price supports plus high demand for commodities during World War II and the Korean War allowed farmers to invest in new equipment and buildings. The continuation of price supports through much of the 1950s, an ongoing labor shortage, and a cost-price squeeze lasting from the mid-1950s until the early 1970s were all powerful influences in farm decision making. Faced with high labor costs and diminishing returns, farmers who hoped to remain in agriculture adopted and adapted new technology and reshaped the land, inaugurating what historian Wayne Rasmussen labeled the second agricultural revolution.[4]

If the period from 1945 to 1972 can be characterized as an agricultural revolution, then farmers were the revolutionaries. They had many allies, one of the most prominent of which was the Cooperative Extension Service. With a network of county directors and experts based in Ames and at regional offices throughout Iowa, extension leaders were in a unique position to influence farmers through periodic contact as well as through formal reports, meetings, and the advice they provided to farm journalists. Farm magazines and newspapers provided forums for the discussion of new techniques, and their writers reported on the varying degree of success farmers were experiencing with new tools and techniques. Manufacturers and advertisers used these magazines and newspapers to reach out to farmers and inform them about extraordinary gains and profits that could be realized with the new technology. Government policies that paid farmers to take land out of cultivation encouraged them to increase profits on the land that remained in production. But for all these voices proclaiming the merits of new technology, none of them could make farmers change their behavior. It was farmers, people with grease under their fingernails and Atrazine and crop oil on their overalls, who industrialized the rural landscape.[5]

The farm landscape of the early 1970s was very different from that of the mid-1940s, even as it remained a place dedicated to agriculture. Vacant houses were not the only symbols of the changes wrought in part by the farmers who industrialized Iowa farms. The new grain bins equipped with stirring devices and crop dryers and the combines, balers, and forage harvesters parked in sheds and barns were also the tools of farm expansion. Automated feeding systems, new dairy parlors, feedlots, pole barns, and confinement feeding operations enabled a handful of farmers to do what had previously taken many hands to accomplish. Corn, soybeans, and hay were now the most important crops, while farmers relegated oats and other small grains to a shrinking percentage of their farms. Fields of thickly planted corn and soybeans were the rule, not the exception, after fertilizer and pesticides became more common and the term "continuous corn" entered farming vocabulary. The sprayers, pesticide containers, and bags of purchased fertilizer and feed were material evidence of the chemicals that allowed farmers to increase both yields per acre and production per animal. By the 1970s new weed and insect species—previously unknown or of minor importance on Iowa farms—now inhabited fields, fencerows, and farmyards because of the technology that farmers were using. These physical manifestations of technological change were signs that Iowa, the heart of the Corn Belt, was an industrial landscape as much as it was a rural one.

As much as farmers enjoyed the many benefits of new technology, they also contended with problems that occupied a growing share of their attention and resources. Chemical techniques created new problems even as they solved old ones. The pesticides farmers used to increase production by limiting populations of competing species allowed other species to thrive. Farmers reduced the threat of broadleaf weeds only to find that grassy-weed species that were resistant to chemicals were now filling the void. Similarly, new insect species thrived when farmers used insecticide to control targeted species. The western corn rootworm moved into the state as farmers reduced populations of the northern corn rootworm. Livestock pests developed resistance or tolerance to certain chemicals when survivors reproduced. The new chemicals that killed resistant species were more toxic to both humans and animals, increasing the dangers of an already hazardous job. In the 1960s farmers applied more fertilizer per acre to maximize yields and employed new kinds of herbicide and in new combinations. Livestock manure was a problem for farmers who concentrated larger herds in feedlots and confinement buildings. Antibiotics appeared to lose some of their growth-enhancing properties during the 1960s, just as pesticides lost potency. Farmers frequently responded by using more chemicals, new formulations, and in some cases they used cultural, biological, or mechanical control techniques.

The ways in which farmers used new technology brought criticism from other farmers, government regulators, and the public. Experts and manufacturers introduced some chemical technology with severe warnings about the toxicity and hazards posed by these chemicals for target species and users. It was only after highly publicized incidents involving urban, suburban, and rural pollution problems that the public became concerned about the consequences of technology use. Concerns about the misuse of chemicals brought unwanted publicity and increasing regulation. Iowa lawmakers began to require training for commercial pesticide applicators, and they regulated livestock feedlots of specified sizes through a system of permits intended to minimize manure runoff. In 1972 the federal government cancelled the registration of DDT and announced a ban of stilbestrol as a feed additive, restricting the use of DES to implants.

Many farmers believed they were being besieged by outsiders who did not understand agriculture or chemical technology. As one journalist noted, farmers were in a "fight for survival."[6]

Farmers of the 1940s would have been surprised at the increasing financial costs of the new technology that by the early 1970s was becoming essential on Iowa's farms. Every farmer lived with the financial consequences of replacing human labor with the machines and chemicals used to maximize production. Figured in constant 1950 dollars, farmers' labor costs fell by over half between 1950 and 1970. Other expenses, however, increased. Power and machinery costs were 30 percent higher while fertilizer expenses more than tripled. Farmers who wanted to stay in agriculture paid these higher costs as a part of doing business. They found that capital expenses for machinery were preferable to increased labor costs. Chemical costs were low in the 1940s and 1950s but steadily increased during the 1960s. Farmers planned to pay for tillage, seed, planting, cultivating, and harvesting, so it made sense to use chemicals to get the largest crop possible out of the investment they were already planning to make.[7]

Farmers who adopted industrial techniques were leaders in the farm-expansion movement of the postwar period. In some cases they did this by performing custom work for other farmers. Custom work was a good solution for those farmers who purchased sprayers, hay balers, and combines to recover some of their investment faster than if they simply used the machines themselves. It was also a way of obtaining labor-saving machinery that might not otherwise be profitable to use on a small farm. Many farmers preferred to expand operations by purchasing or renting land, rather than doing custom work, in order to get full benefits of their investment in technology. A contemporary study of Iowa farm records for the 1968 season indicated that those who farmed more than 500 acres paid much less in operating cost than smaller farms, with the biggest saving for machinery and fuel. People who farmed 160 acres paid $32 per acre for machinery, power, and fuel, while those who farmed 320 acres paid $25 per acre, and those who farmed 600 acres paid only $19 per acre. The difference in overall farm costs per acre was even more remarkable. The family with 160 acres paid $107 per acre in operating expenses and labor, while the farmer with 600 acres paid only $64 per acre. The logic of industrial agriculture rewarded those who were able to do what President Nixon's secretary of agriculture, Earl Butz, counseled: get bigger or get better.[8]

Getting out of farming altogether was Butz's third suggestion, and many farm families heeded it. Many farmers left agriculture for other employment or for retirement, but the land they left behind stayed in production. With fewer farmers there was more land per farm, especially after 1950 when the rate of migration from farm to town accelerated. The size of the average Iowa farm in 1945 was approximately 169 acres, but by 1970 the average farm was 249 acres. Farmers who owned all the land they farmed increased the size of their landholdings only a little, however. Land values increased with demand and made it more difficult for potential purchasers. The most common response was to rent rather than buy. This strategy, known as part-ownership, allowed farmers to use their new equipment to increase the scale of operations without going into debt for land purchases. Farm families who were part-owners increased the size of their operations from 205 acres in 1949 to over 300 acres in 1970. The proportion of farmland worked by part-owners in Iowa and the other Corn Belt states of Illinois, Indiana, and Ohio increased from under 20 percent in 1940 to approximately 45 percent in 1970. Simultaneously, the proportion of land farmed by full owners declined from approximately half to less than 40 percent. Expansion was here to stay, and for most farm families it involved a combination of ownership and renting.[9]

The industrialization of Corn Belt agriculture allowed many farmers to expand and maximize production, but the very success of the industrial ideal was unsettling to some Iowa producers. A 1971 survey of Iowa farmers indicated that 69 percent believed pollution was not yet a serious problem, although they conceded that measures for prevention were necessary. Another 20 percent believed pollution was a serious problem and that a vigorous control program was needed. It is not surprising that, for many farmers, pollution and damage to ecosystems were other producers' problems. There were, however, people who believed in reforming agriculture from the ground up.[10]

A minority of farmers and some would-be farmers concerned about the environmental hazards of agriculture and food safety searched for alternate production strategies. In the 1970s a handful of farmers embraced an organic ethos, accepting many of the mechanical aspects of the productive revolution but rejecting most of the chemical techniques. In Cass County, Denise O'Brien, who began organic farming in 1976 with her husband, Larry Harris, recalled that some of those who rebelled against industrialization did so for immediate reasons more so than for abstract concerns about ecosystems. Several of the pioneering Iowa organic farmers whom O'Brien and Harris met in the mid-1970s turned to organic techniques because they knew someone who had experienced injury from farm chemical use. By the early twenty-first century, organic farming had become one of the most profitable sectors of the agricultural economy, with some crops such as organic soybeans commanding significant premiums in the marketplace. Organic farmers drew on much of the agricultural system that industrial production had displaced, notably crop rotations that include small grains and forage crops that check pest populations and lessen the draw on soil nutrients. By 2000 the USDA had certified only a small portion of Iowa's farm production as organic. But this small and growing counterculture movement is the exception that proves the rule of agricultural industrialization in the Corn Belt, highlighting the extent to which farmers used industrial techniques to transform farm landscape in the postwar period.[11]

Notes

Introduction

1. Jules B. Hilliard, "The Revolution in American Agriculture," *National Geographic* (February 1970): 147–85. This definition of the industrialization of agriculture is from Deborah Fitzgerald, *Every Farm a Factory: The Industrial Ideal in American Agriculture* (New Haven: Yale University Press, 2003), 22–23.

2. Sam Bowers Hilliard, "The Dynamics of Power: Recent Trends in Mechanization on the American Farm" *Technology and Culture* 13 (January 1972): 1–24; Wayne D. Rasmussen, "The Impact of Technological Change on American Agriculture, 1862–1962," *Journal of Economic History* 4 (December 1962); John T. Schlebecker, *Whereby We Thrive: A History of American Farming, 1607–1972* (Ames: Iowa State University Press, 1975); John Shover, *First Majority, Last Minority: The Transformation of Rural Life in America* (DeKalb: Northern Illinois University Press, 1977), xvi, 143.

3. *Forty-Sixth Annual Iowa Year Book of Agriculture, 1945* (Des Moines: State of Iowa, 1945), 737; *Tenth Biennial Report of Iowa Year Book of Agriculture, 1970–1971* (Des Moines: State of Iowa, 1972), 344.

4. John C. Hudson, *Making the Corn Belt: A Geographical History of Middle-Western Agriculture* (Bloomington and Indianapolis: Indiana University Press, 1994), chs. 7, 9, 10.

5. Ladd Haystead and Gilbert C. Fite, *The Agricultural Regions of the United States* (Norman: University of Oklahoma Press, 1955), 144, 149.

6. Allan G. Bogue argued that important changes took place in production in the Corn Belt between 1900 and 1939, including the adoption of the tractor, hybrid seed corn, and the mechanical corn picker. These changes were important, but farming in 1940 looked so much like farming in 1900 that a person who successfully operated a farm in 1900 could successfully operate the same farm in 1940 without significantly changing management or production strategies. Gilbert Fite observed that farmers on South Dakota farms utilized horse-drawn equipment with tractors throughout much of the interwar period. Only in the 1930s did manufacturers produce large numbers and a wide variety of implements adapted for tractor use, limiting the ways in which tractors could be used. Allan G. Bogue, "Changes in Mechanical and Plant Technology: The Corn Belt, 1910–1940," *Journal of Economic History* 43 (March 1983): 2; Gilbert Fite, "The Transformation of South Dakota Agriculture: The Effects of Mechanization, 1939–1964," *South Dakota History* 19 (Fall 1989): 278–83; also R. Douglas Hurt, "Ohio Agriculture since World War II," *Ohio History* 97 (Winter–Spring 1988): 50–71.

7. Mary Neth detailed the ways midwestern farm women "made do" to preserve their farms, while Pamela Riney-Kehrberg documented how Kansas families managed to stay on farms and in small towns in the Dust Bowl. Katherine Jellison showed how midwestern women used technology to reinforce their roles as family producers rather than consumers during these years. Alan I Marcus and Howard P. Segal described the mood of post–World War II optimism about technology. Mary Neth, *Preserving the*

198 NOTES TO PAGES 8–10

Family Farm: Women, Community, and the Foundations of Agribusiness in the Midwest, 1900–1940 (Baltimore: Johns Hopkins University Press, 1995); Pamela Riney-Kehrberg, *Rooted in Dust: Surviving Drought and Depression in Southwestern Kansas* (Lawrence: University Press of Kansas, 1994); Katherine Jellison, *Entitled to Power: Farm Women and Technology, 1913–1963* (Chapel Hill: University of North Carolina Press, 1993); Alan I Marcus and Howard P. Segal, *Technology in America: A Brief History* (San Diego: Harcourt, Brace, Jovanovich, 1989), 309–11.

8. Rasmussen labeled this period the second agricultural revolution ("Technological Change," 588). Changes in agricultural production occurred worldwide as farmers in industrialized nations applied industrial ideals to agriculture. For a comparison of how this process occurred in the German Federal Republic and German Democratic Republic during the same time period, see Arnd Bauerkamper, "The Industrialization of Agriculture and Its Consequences for the Natural Environment: An Inter-German Comparative Perspective," *Historical Social Research*, ed. Frank Uekotter, 29.3 (2004): 124–49. The author is grateful to Dr. Uekotter for calling this article to his attention.

9. Wartime labor shortages were acute in the Midwest, with the states of Iowa, Kansas, Missouri, Minnesota, Nebraska, and North and South Dakota averaging a loss of 27 percent of hired workers from 1939 to 1945. Rural depopulation did not slow down after the war, with both farm operators and hired men either supplementing farmwork with city employment or leaving farming altogether. Walter W. Wilcox, *The Farmer in the Second World War* (Ames: Iowa State College Press, 1947), 100; *Forty-Sixth Annual Iowa Year Book of Agriculture* (Des Moines: State of Iowa, 1945), 15–16; Paul J. Jehlik, "Iowa Farmers Using Less Hired Help," *Iowa Farm Science* 6 (June 1952); John D. Hervey, "A Kingdom for a Hired Hand," *Successful Farming* (March 1946); Paul R. Robbins, "Labor Situations Facing the Producer," in *The Pork Industry: Problems and Progress*, ed. David G. Topel (Ames: Iowa State University Press, 1968). Donald Holley's study of the role played by the mechanical cotton picker and by African American migration in shaping the South is a good example of how labor shortages and associated increases in wage rates helped convince farmers to use machines. *The Second Great Emancipation: The Mechanical Cotton Picker, Black Migration, and How They Shaped the Modern South* (Fayetteville: University of Arkansas Press, 2000).

10. "What Happens to Rural High School Graduates?" *Wallaces Farmer and Iowa Homestead*, 16 February 1957; *U.S. Census of Population, 1960. Volume I, Characteristics of the Population, Part 17, Iowa* (Washington, DC: GPO, 1963), 173; Bruce L. Gardner, *American Agriculture in the Twentieth Century: How It Flourished and What It Cost* (Cambridge: Harvard University Press, 2002), 172–74. *Wallaces Farmer* changed names several times during the study period, from *Wallaces' Farmer and Iowa Homestead* to *Wallaces' Farmer* and finally to *Wallaces Farmer*. I will use the latter citation throughout.

11. Gilbert C. Fite, *American Farmers: The New Minority* (Bloomington: Indiana University Press, 1981), 107, 118–19. Richard G. Bremer argued that after World War II farmers in the Loup River country of Nebraska used technology to manage the risks of farming in turbulent times. *Agricultural Change in an Urban Age: The Loup Country of Nebraska, 1910–1970* (Lincoln: University of Nebraska, 1976), 202. Dick Albrecht, "What's Happening to Farm Income?" *Wallaces Farmer*, 2 January 1960.

12. Allan G. Bogue, *From Prairie to Corn Belt: Farming on the Illinois and Iowa Prairies in the Nineteenth Century* (Chicago: University of Chicago Press, 1963), 1; Robert C. McMath, "Where's the 'Culture' in Agricultural Technology?" in *Outstanding in His Field: Perspectives on American Agriculture in Honor of Wayne D. Rasmussen*, ed. Frederick V. Carstensen, Morton Rothstein, and Joseph A. Swanson (Ames: Iowa State University Press, 1993), 125. Without denying the importance of chemical manufacturers and extension professionals in the process of technological adoption, the current study brings the users to the forefront. For Iowa farmers during the postwar years, herbicide was a product that would help them maximize production and re-

duce labor. As the following pages demonstrate, the fact that farmers used technology in ways not prescribed by manufacturers and experts suggests that they domesticated technology in a manner described by Nelly Oudshoorn and Trevor Pitch in *How Users Matter: The Construction of Users and Technologies* (Cambridge: MIT Press, 2003), 14–15.

13. "How Farm People Accept New Ideas," *Special Report No. 15,* Iowa State College, Ames, 1955; Joe M. Bohlen, "Adoption and Diffusion of Ideas in Agriculture," in *Our Changing Rural Society,* ed. James H. Copp (Ames: Iowa State University Press, 1964); George M. Beal and Everett M. Rogers, "The Adoption of Two Farm Practices in a Central Iowa Community," *Special Report No. 26,* Agricultural and Home Economics Experiment Station, Iowa State University, Ames, June 1960. Historian Mark R. Finlay observed in his study of postwar hog production that not all farmers adopted a "Fordist" approach to agriculture. See "Hogs, Antibiotics, and the Industrial Environments of Postwar Agriculture," in *Industrializing Organisms: Introducing Evolutionary History,* ed. Susan R. Schrepfer and Philip Scranton (New York and London: Routledge, 2004), 237–60.

14. Kenneth L. Peoples, David Freshwater, Gregory D. Hanson, Paul T. Prentice, Eric P. Thor, *Anatomy of an American Agricultural Credit Crisis: Farm Debt in the 1980s* (Lanham, MD: Rowan and Littlefield, 1992), 3–5.

15. Lowell Soike observed the differences between contemporary and historic farm landscapes in the introduction to his essay "Viewing Iowa Farmsteads," in *Take This Exit: Rediscovering the Iowa Landscape,* ed. Robert F. Sayre (Ames: Iowa State University Press, 1989), 154.

1—Insecticide: Time for Action

1. "Borer Racing across Iowa," *Wallaces Farmer,* 21 August 1943.

2. "Standard 25 Percent DDT Concentrate," advertisement, *Wallaces Farmer,* 7 June 1952.

3. *Wallaces Farmer* articles "Borer Racing across Iowa," 21 August 1943, and "Corn Borer Damage over 4 Per Cent," 14 April 1944; *Annual Report, Entomology, 1945,* Summary, 3 (quote). The *Annual Report for Entomology* (later titled *Annual Report for Entomology and Wildlife*) is among the most valuable sources used in the preparation of this manuscript and deserves some explanation. The yearly reports include a wide variety of information, including summaries of work performed by Iowa State College (ISC renamed ISU in 1959) extension entomologists, project reports, correspondence to and from the state entomologists, memos to county agents, technical reports of chemical- and cultural-control experiments from cooperating farmers, radio and television scripts and program outlines, texts of talks given to youth groups, surveys of insecticide use, photographs, programs from training meetings and conferences, newspaper clippings, and recommendations for insecticide use. Only a few years of these reports have consistent page numbers throughout the entire bound volume; most of them are not organized as a unit. In citing information from this diverse and invaluable source, the author has provided as much specific information as possible about the particular evidence from within the volume. All of the reports will be cited as *Annual Report, Entomology,* accompanied by other identifying information and the year of the report. The collection is located at Special Collections, Parks Library, Iowa State University, and is listed as "Records, Cooperative Extension Service, Entomology and Wildlife."

4. *Annual Report, Entomology, 1946,* Summary, 1; "Kill Borers with DDT Dust," *Wallaces Farmer,* 1 September 1945 (quote).

5. "Corn Borer Control, Discussion for 4-H Club Meetings," *Annual Report, Entomology, 1945;* ibid., 1946, 1–2; Harold Gunderson to County Extension Directors, 18 February 1947, in ibid., 1947.

6. J. G. Gunning, "Plow to Kill the Borer," *Successful Farming* (March 1946); *Annual Report, Entomology, 1946*, 1–2 (quote).

7. *Wallaces Farmer* articles "DDT Reduces Borer Loss," 21 February 1948, "Use DDT to Kill Corn Borers," 7 February 1948, "Old Ways Failed to Stop Corn Borers," 17 April 1948 (quotes), and "Keep Fighting Borers," 20 November 1948 (quote). O. T. Zimmerman and Irvin Lavine, *DDT: Killer of Killers* (Dover, NH: Industrial Research Service, 1946), is a good example of the kind of promotional material of the late 1940s.

8. Harold Gunderson, "Stop Corn Borer Damage!" *Pamphlet 134* (May 1948), and "Controlling the European Corn Borer," *Pamphlet 150* (December 1949), both for the Agricultural Extension Service, Iowa State College, Ames; Harold Gunderson and E. P. Sylwester, "Spraying—Babyhood to Manhood in 5 Years," E. P. Sylwester Papers, Special Collections, Parks Library, Iowa State University; "Sprayer Market in '50 Looks Good," *Farm Implement News*, 10 February 1950.

9. Wally Inman, "Ready for the Borers?" *Wallaces Farmer*, 3 June 1950; Farm Radio Service, 11 August 1947, *Annual Report, Entomology, 1947* (quote).

10. Yearly statistics for the estimated use of chemicals for fly, European corn borer, and corn rootworms are from the *Annual Report, Entomology* summaries (figure for total corn acreage is from 1949). USDA *Agricultural Statistics 1950* (Washington, DC: GPO, 1950), 46. At least three versions of a corn-borer control pamphlet were in circulation before 1970: "The European Corn Borer and Its Control in the North Central States," *Pamphlet 176*, Cooperative Extension Service and Agricultural and Home Economics Experiment Station, Iowa State University, Ames, revised in October 1961 and October 1968.

11. USDA researchers worked on other methods of control for the European corn borer also, including the release of 154,000 wasplike parasites in fourteen Iowa counties in 1944, attempting to replicate earlier success with other insects in California. Richard C. Sawyer, "Monopolizing the Insect Trade: Biological Control in the USDA, 1888–1951," in *The United States Department of Agriculture in Historical Perspective*, ed. Alan I Marcus and Richard Lowitt (Washington, DC: Agricultural History Society, 1991): 271–85. In 1959 farmers read about the development of a "living insecticide." The bacteria *bacillus thuringiensis* (bt) produces a spore within the borer larvae that triggers a disease that is lethal to the borer. This bt corn became a significant technique for controlling European corn borers in the late twentieth and early twenty-first centuries, as scientists developed technology to modify seed plants genetically. *Wallaces Farmer* articles "New Living Insecticide to Go on Field Trials," 7 February 1959, and Keith Remy, "A Biological Control for the Corn Borer?" 7 March 1964.

12. European Corn Borer Conference, Cedar Rapids, Iowa, 16–17 January 1946, *Annual Report, Entomology, 1946* (panel of farmers); Keith E. Myers to Harold Gunderson, 2 March 1946, ibid. (quote); Keith Remy, "What's Happened to the Corn Borer?" *Wallaces Farmer*, 1 February 1964 (Brindley).

13. Research and breeding efforts are described in a promotional comic book depicting a father and son visiting the Northrup King Research farm at Shakopee, Minnesota. "KX [Kingscrost] Research at Work for You," KX-57-4, Northrup, King & Company, 1957; Remy, "What's Happened to the Corn Borer?" *Wallaces Farmer*, 1 February 1964 (quote).

14. Fremont Conrad, "Resistant Corn Checks Borer," *Successful Farming* (November 1942); Dennis Havran, interview with author, 11 January 2002, Des Moines, tape recording; Bob Nymand, interview with author, 30 October 2002, Brayton, Iowa, tape recording.

15. Papers of Rudolf E. Schipull, Joseph Ludwig, and William M. Adams, all in Special Collections, Parks Library, Iowa State University.

16. D. Ivan Johannes to Harold Gunderson, 4 May 1959, *Annual Report, Entomology, 1959*.

17. "Farmers Strong for DDT," *Wallaces Farmer,* 17 April 1948 (quote, p. 12); *Annual Report, Entomology, 1947,* Summary, p. 9; Pestroy advertisement, *Wallaces Farmer,* 7 June 1947; "New Chemicals Mean Death to Pests," *Wallaces Farmer,* 7 August 1948 (quote).

18. *Wallaces Farmer* articles "DDT Is Slow Poison," 1 September 1945 (August), "DDT Raises Summer Milk Production," 3 August 1946, and "Farmers Strong for DDT," 17 April 1948 (quotes, p. 12).

19. "Declare War on Flies," *Wallaces Farmer,* 15 March 1947; *Annual Report, Entomology, 1949.*

20. "Iowa Farms Can Be Made Fly-Free," *Wallaces Farmer,* 7 May 1949.

21. J. J. Davis, "A Word on DDT," *Successful Farming* (April 1946); Jim Roe, "The Latest on DDT," *Successful Farming* (June 1946); "Declare War on Flies," *Wallaces Farmer,* 15 March 1947; Earle S. Raun, "Flies on Livestock," *Pamphlet 200,* Agricultural Extension Service, Iowa State College, Ames, May 1953; "FDA to Bear Down on Chemical Misuse," *Wallaces Farmer,* 7 May 1960 (quote, p. 6).

22. *Wallaces Farmer* articles "Iowa Farms Can Be Made Fly-Free," 7 May 1949, "Clean Up to Get Rid of Flies," 15 April 1950 (quote, p. 34), "Get Early Start on Fly Control," 17 March 1951 (quote), "No Need to Live with Flies," 3 May 1952, "You Can't Afford to Feed Flies," 20 June 1953, and "Does Fly Control Work for You?" 19 June 1954.

23. "You Can't Afford to Feed Flies," *Wallaces Farmer,* 20 June 1953.

24. "Fly Spray Survey," *Annual Report, Entomology, 1951;* "Sprayed Cows with Lindane," *Wallaces Farmer,* 17 July 1954.

25. Earle S. Raun, "Kill Flies This Summer," *Successful Farming* (July 1959); *Midwest Farm Handbook,* 5th ed. (Ames: Iowa State University Press, 1960), 308–10; *Wallaces Farmer* articles "Insecticides Aid in Controlling Flies," 7 June 1958 (quote, p. 18), "Insect Experts Talk It Over," 16 April 1955, and "New Chemical for Flies," 20 March 1954.

26. Extension entomologists provided estimates in each year's annual report of the number of farmers using chemicals based on county agent reports. Specific fly-control strategies are detailed in "Flies on Livestock," *Pamphlet 200,* Agricultural Extension Service, Iowa State College, Ames, May 1953.

27. Schipull Papers, Box 1, Folder 14, Box 2, Folders 2–4, 6; Ludwig Papers, 2 July 1947, 3 July 1954, 3 January 1955, 20 February 1956; and Adams Papers, 23 June 1948, 24 June 1950.

28. "To All County Extension Personnel," *Annual Report, Entomology, 1960* (Gunderson); *Annual Report, County Extension Activities, Clinton County, 1961,* 19 (county extension activities annual reports are hereafter cited as county name annual report, followed by date and page number); *Fayette County Annual Report 1965,* 6; Bob Dunaway, "Keep Insecticides out of Milk," *Wallaces Farmer,* 13 August 1966.

29. *Annual Report, Entomology, 1952,* 16; Scotty Woods, "Now Granules Control Soil Insects," *Successful Farming* (April 1959).

30. *Annual Report, Entomology, 1953,* 9; J. H. Lilly and Harold Gunderson, "Fighting the Corn Rootworm," *Iowa Farm Science* 6 (February 1952); "Feed Insects or Fight 'Em," *Wallaces Farmer,* 5 March 1955 (quote); Joe M. Bohlen, George M. Beal, and Daryl Hobbs, "The Iowa Farmer and Farm Chemicals: Attitudes, Level of Knowledge and Patterns of Use," *Rural Sociology Report No. 3,* Department of Economics and Sociology, ISU, Ames, November 1958, 12; "Get the Jump on Rootworms!" *Wallaces Farmer,* 2 April 1955; Schipull Papers; Ludwig Papers.

31. Harold Gunderson and J. H. Lilly, "Control Corn Rootworms," *Pamphlet 178,* Agricultural Extension Service, Iowa State College, Ames, revised April 1953.

32. "Rootworm Control for 1962," *Wallaces Farmer,* 7 April 1962; "Proceedings —Corn Rootworm Conference," *Annual Report, Entomology, 1963;* "Resistant Corn Rootworms Now in All Iowa Counties," *Wallaces Farmer,* 5 February 1966.

33. "It Pays to Control Soil Insects," *Wallaces Farmer,* 6 April 1963; "Control Recommendations for Corn Rootworm Announced, Iowa State University Information Service, for Release Saturday, November 9, 1963," *Annual Report, Entomology, 1963.* In *Wallaces Farmer,* Ken Hofmeyer, "What Farmers Plan to Do for Corn Rootworms," 18 April 1964 (quote), American Cyanamid Company advertisement, 6 February 1965, and Keith Remy, "Ways to Control Soil Insects in Your 1964 Cornfields," 21 March 1964 (quote). Lloyd E. Zeman, "How to Cope with New Corn Rootworm Problems," *Successful Farming* (January 1964).

34. "Insect Information Letter No. 5," 17 May 1965, *Annual Report, Entomology, 1965* (Gunderson); *Wallaces Farmer* articles Ron Lutz, "Careless Pesticide Handling Causing Livestock Death Losses," 14 March 1970, Ken Hofmeyer, "Play It Safe with Rootworm Chemicals," 27 April 1968, "Pesticide Accidents," 13 January 1968, "Protect Yourself from Chemicals," 26 April 1969, and Monte Sesker, "Use Chemicals Safely!" 11 April 1970.

35. "Insect Information Letter No. 4," 4 May 1964, *Annual Report, Entomology, 1964.*

36. "Pesticide Residue Discussion," 2 November 1965, *Annual Report, Entomology, 1965; Seventh Biennial Report of Iowa Book of Agriculture, 1964–1965* (Des Moines: State of Iowa, 1966), 71–72.

37. "License Applicators of Persistent Pesticides," *Wallaces Farmer,* 11 September 1971.

38. Robert L. Morris and Lauren G. Johnson, "Pollution Problems in Iowa," in *Water Resources of Iowa,* ed. Paul J. Horick (Iowa City: University Printing Service, 1970), 98, 100, 107.

39. *Tenth Biennial Report of Iowa Year Book of Agriculture, 1970–1971* (Des Moines: State of Iowa, 1972), 81–83.

40. Charlie Nettles, "More Confusion over Pesticides," *Des Moines Sunday Register,* 29 November 1970.

2—Herbicide versus Weedy the Thief

1. Dr. Bob Hartzler, Iowa State University extension weed management specialist, found a copy of the skit in some files in his office and posted it on his Web site www.weeds.iastate.edu/weednews/24dplay.htm. The script is undated, but it was probably written sometime between 1945 and 1947, since 2,4-D was released commercially in 1945 and its author, Pearl Converse, died on 4 May 1947. *Ames Daily Tribune,* 5 May 1947.

2. The commercial introduction of 2,4-D occurred during a period of optimism about technology. See Marcus and Segal, *Technology in America,* ch. 7; also Edmund Russell, *War and Nature,* ch. 9. An excellent example of faith in chemical technology for solving pest problems can be seen in Joshua Blu Buhs, *The Fire Ant Wars: Nature, Science, and Public Policy in Twentieth Century America* (Chicago: University of Chicago Press, 2004).

3. Nicolas Rasmussen, "Plant Hormones in War and Peace: Science, Industry, and Government in the Development of Herbicides in 1940s America," *Isis* 92 (June 2001): 309–11; Gale Petersen, "The Discovery and Development of 2,4-D," *Agricultural History* 41 (July 1967): 243–53; Orvin C. Burnside, "The History of 2,4-D and Its Impact on Development of the Discipline of Weed Science in the United States," in Burnside, *Phenoxy Herbicides,* 5–11; L. W. Kephart, "Weed Control with Chemicals," *Agricultural Engineering* (November 1946): 506–11; John W. Mitchell, "Plant Growth Regulators," in *Science in Farming: The Yearbook of Agriculture, 1943–1947* (Washington, DC: GPO, 1947): 256–66; *Wallaces Farmer* articles "Trying New Kind of Weed-Killer," 3 March 1945, and "New Weed Sprays," 17 November 1945.

4. "Weeds Won in War Years," *Wallaces Farmer,* 18 January 1947; *Hamilton County Annual Report, 1946,* 7 (director); *Humboldt County Annual Report, 1945,* 10; Clifford Sams, interview by author, 13 September 2002, Conrad, Iowa, tape recording.

5. *Wallaces Farmer* articles "Weed Law Has Teeth," 7 June 1947, "New Law Helping in Weed Control," 6 November 1948, and "Winning Weed Fight," 18 June 1949.

6. *Humboldt County Annual Report, 1945,* 10; *Hardin County Annual Report, 1945,* 6; Annual Report Summary—Weed Control, 1 October 1947–30 September 1948, Sylwester Papers; *Fiftieth Annual Iowa Year Book of Agriculture* (Des Moines: State of Iowa, 1949), 451; *Wallaces Farmer* articles "Coming Events," 1 January 1949, "College Sets Field Days," 19 June 1948, and "Four Iowa State College Farm Clinics," 2 February 1950.

7. E. P. Sylwester, A. L. Bakke, and D. W. Staniforth, "Recommendations for Chemical Weed Control," *Pamphlet 140,* Agricultural Extension Service, Iowa State College, March 1949 (quote). *Wallaces Farmer* articles "Here's How to Kill Weeds," 19 March 1949 (quote), "Don't Plant Weed Seeds in 1950," 3 March 1950, "Don't Let Weeds Whip You," 2 February 1952, "Time Cultivation to Kill Weeds," 7 June 1952, "Get Corn Weeds with First Cultivation," 16 May 1953, and "Don't Throw away Your Cultivator," 7 May 1960, "Weeds 2,4-D Will Kill," 1 January 1949; *Wallaces Farmer* articles by Monte Sesker, "Chemicals and Cultivation Control Weeds in Corn," 27 May 1967, and "Stop Those Weeds in Your Cornfield," 10 May 1969; Gene Neven, interview by author, 12 October 2002, Marshalltown, Iowa, tape recording; L. W. Kephart, "2,4-D, New Killer for Weeds," *Successful Farming* (March 1946); "Weed Control in Corn," *Pamphlet 269,* Iowa State University Extension, March 1964; "Weed Control in Soybeans," *Pamphlet 270,* Iowa State University Extension, April 1969.

8. Jim Roe, "We Used 2,4-D on Corn," *Successful Farming* (May 1949) (quote, p. 40); *Wallaces Farmer* articles "Bad News for Weeds," 1 January 1949 (quote) and Wally Inman, "Cheaper Weed Kills," 20 August 1949 (Erickson, p. 8).

9. *Wallaces Farmer* articles "Poor Future for Weeds," 3 June 1950, "Farmers Kill Weeds with Chemicals," 5 March 1949, "Kill Corn Weeds with 2,4-D?" 17 April 1954, and "Control Iowa's Problem Weeds," 18 May 1963; Bohlen, Beal, and Hobbs, "Iowa Farmer and Farm Chemicals," 9.

10. Sams interview.

11. Gunderson and Sylwester, "Spraying—Babyhood to Manhood in 5 Years," Sylwester Papers; "Sprayer Market in '50 Looks Good," *Farm Implement News,* 10 February 1950.

12. *Wallaces Farmer* articles "Poor Future for Weeds," 3 June 1950, and "How Does 2,4-D Work for You?" 18 June 1955; Bohlen, Beal, and Hobbs, "Iowa Farmer and Farm Chemicals," 11.

13. Dana Stewart, "What 10,000 Acres of Weed Spraying Taught Me," *Successful Farming* (June 1954); *Grundy County Annual Report, 1948,* "Information on Weed Control Program," 12.

14. *Hancock County Annual Report, 1952,* 5; *Grundy County Annual Report, 1957,* 4 (quote).

15. *Hamilton County Annual Report, 1947,* 6–7.

16. *West Pottawattamie County Annual Report, 1950,* 6, "Information on Weed Control Program," 13; *West Pottawattamie County Annual Report, 1951,* 8–9, "Information on Weed Control Program," 12; *West Pottawattamie County Annual Report, 1952,* 5, "Information on Weed Control Program," 11; *West Pottawattamie County Annual Report, 1953,* 4, "Information on Weed Control Program," 11.

17. J. Clifford Johnson, interview by author, 14 December 2002, Council Bluffs, Iowa, tape recording; A. F. Martin to Dr. E. P. Sylwester, 19 September 1952, and "Procedures to Help Minimize Danger from Spray 'Fumes' and Spray 'Drift' in the Council Bluffs Area," both in Sylwester Papers; Fred Slife, "The Most Asked Questions about Basal Spray," *Successful Farming* (February 1954).

18. Iowa fruit growers were dealing with other problems that were possibly even more lethal for their business than herbicide. From 1943 until 1964, the Bracero

program provided a government-subsidized labor force for Pacific Coast growers that allowed them to reduce or hold the line on labor costs. This was a luxury Iowa farmers did not enjoy. C. E. White to H. E. Rea, 6 February 1955, Sylwester Papers, box 9, folder 4; *West Pottawattamie County Annual Report, 1954, 1955, 1956, 1957;* Cecil J. Baxter to E. P. Sylwester, 10 June 1961, and Ed Heins, "Urge State Ban on 2,4-D to Save Fruit," *Des Moines Register,* no date, both in Sylwester Papers, box 9, folder 4; Neven interview.

19. "2,4-D Kills Weeds, not Livestock," *Successful Farming* (June 1959); *Wallaces Farmer* articles "Workday Pointers," for 1 June 1946, 5 October 1946, and 20 August 1949, "2,4-D Not Harmful to Livestock," 7 June 1947, "Confused on Uses of 2,4-D and DDT," 21 June 1947, and "2,4-D Won't Kill Cows," 19 August 1950. The authors of a 1996 review of health risks associated with phenoxy herbicides such as 2,4-D concluded that the public is not at risk of any health problems from 2,4-D, but people who work with the chemical are exposed to potentially harmful doses and should wear protective equipment to minimize risk. See Rebecca A. Johnson and Elizabeth V. Wattenberg, "Risk Assessment of Phenoxy Herbicides: An Overview of the Epidemiology and Toxicology Data," in Burnside, *Phenoxy Herbicides;* Nymand interview; *Holladay v. Chicago, Burlington & Quincy Railroad Company,* 255 F. Supp. 879 (U.S. Dist. 1966).

20. Ludwig Papers; Adams Papers.

21. *Wallaces Farmer* advertisements, for Essick and Speedy Sprayer, 19 March 1949, for Kim's Fast-O-Matic, 15 March 1958, for Century, 3 May 1958; Adams Papers; Havran interview; Flora and Ferd Jarrott Papers, State Historical Society of Iowa, Iowa City.

22. "Recommendations for Chemical Weed Control," *Pamphlet 140,* Iowa State Extension Service, March 1949; Ludwig Papers.

23. *Wallaces Farmer* articles "Poor Year for 2,4-D Weed Test," 17 July 1948, "Approve only Three Bean Varieties," 2 October 1948, Wally Inman, "2,4-D Rules for '49," 15 January 1949, "Seedbed Preparation Helps Kill Weeds," 1 April 1950, Wally Inman, "Kill Weeds in '51," 6 January 1951 (quote), and "Kill Weeds with Pre-Emergence?" 17 March 1953.

24. *Humboldt County Annual Report, 1958,* 13 (extension director); *Wallaces Farmer* articles "How Pre-Emergence Herbicides Work," and Richard Hagen, "Iowa Farm Report on Pre-Emergence Herbicides," both 1 July 1961.

25. *Wallaces Farmer* articles "Need Weed-Killing Help?" 19 May 1951, "Cultivate Corn and Beans," 6 June 1953 (Fridley), and "Cultivate or Spray?" 19 May 1956; Stewart, "What 10,000 Acres of Spraying Taught Me," *Successful Farming* (June 1954).

26. *Clay County Annual Report, 1953,* 4; "Clean Seedbed Will Help Control Weeds," *Wallaces Farmer,* 3 May 1958 (Sylwester); *Hardin County Annual Report, 1961,* 25.

27. James A. Young, "The Public Response to the Catastrophic Spread of Russian Thistle (1880) and Halogeton (1945)," in *Publicly Sponsored Agricultural Research in the United States: Past, Present and Future,* ed. David B. Danbom (Washington, DC: Agricultural History Society, 1988), 122–30. Mark Fiege argued that "nature changes what humans build, often in unanticipated ways; sometimes nature comes back more powerful than before," an insight that informs the story of Iowa farmers' efforts to control nature through chemicals. Mark Fiege, *Irrigated Eden: The Making of an Agricultural Landscape in the American West* (Seattle: University of Washington Press, 1999), 9. Hadley Read, "New Weed Threatens Cornbelt," *Successful Farming* (June 1951); *Wallaces Farmer* articles "Grassy Weeds Threaten," 2 January 1954 (quotes), "Learn More about Weed Control," 1 January 1955, "Grassy Weeds Threaten Iowa Cornfields," 26 April 1969, and "Chemical Control for Smartweed in Beans," 25 April 1970; *Greene County Annual Report, 1956,* 1–2; Fred W. Slife and Lloyd E. Zeman, "1965 Weed and Insect Control Guide," *Successful Farming* (March 1965); Nymand interview.

28. "What Can You Do about Foxtail?" *Wallaces Farmer*, 18 February 1961.

29. Schipull Papers.

30. Ludwig Papers.

31. Charles E. Sommers and Ellery L. Knake, "1969 Corn-Soybean Herbicide Selection Guide," *Successful Farming* (February 1969); *Hardin County Annual Report, 1960*, 25; Richard Krumme, Phil Jones, and Charles E. Sommers, "Your Nine Big Corn Decisions," *Successful Farming* (November 1970).

32. *Henry County Annual Report, 1966*, 16–17; *Hamilton County Annual Report, 1967*, 9; Ken Hofmeyer, "Weed Control in Corn and Soybeans," *Wallaces Farmer*, 10 February 1968.

33. Geigy/Atrazine advertisement, *Successful Farming* (March 1968); "Know More about Your Corn Crop," *Wallaces Farmer*, 11 March 1967; *Grundy County Annual Report, 1965*, 3.

34. R. H. Beatty to D. W. Staniforth, 23 May 1966, M. B. Turner to R. H. Beatty, 17 May 1966, David W. Staniforth Papers, Box 2, Folder 31, Special Collections, Parks Library, Iowa State University.

35. "Dear County Agent" circular, 7 April 1965, Staniforth Papers; "New Dimension in Weed Control," Elanco Products Company, Indianapolis, Indiana, brochure EA 5065; "Incorporating Herbicides—Helpful with Some," *Successful Farming* (March 1968); *Wallaces Farmer* articles Keith Remy, "Which Herbicides Should Be Incorporated into the Soil?" 9 April 1966, "Work Herbicides into the Soil?" 8 April 1967, and "Incorporating Chemicals," 9 May 1970.

36. *Wallaces Farmer* articles "Are Two Herbicides Better than One?" 5 March 1966 (quote, p. 40), "Herbicide Mixtures for Weed Control," 8 April 1967, and "Herbicide Mixtures Offer Some Benefits," 12 April 1969 (quote, p. 85); "Piggyback Sprays: New Way to Clean Out Weeds in Soybeans," *Farm Journal* (April 1971).

37. *Wallaces Farmer* articles "How Surfactants Improve Weed Killers," 9 March 1968, and "Timing Critical When Applying Atrazine-Oil," 10 May 1969.

38. Wally Inman, "2,4-D Rules for '49," *Wallaces Farmer*, 15 January 1949; "Weed Control for Corn and Soybeans," *Successful Farming* (February 1969) (quote, p. 64); "Fit Chemical to Crop Rotation," *Wallaces Farmer*, 8 March 1969; EPA Hazardous Materials Advisory Committee, *Herbicide Report: Chemistry and Analysis, Environmental Effects, Agricultural and Other Applied Uses* (Washington, DC: GPO, May 1974), 58.

39. Untitled editorial, *Wallaces Farmer*, 21 May 1949; Roy J. Reiman and Lloyd E. Zeman, "These Farmers Shoot the Works on Corn," *Successful Farming* (March 1965) (Gabeline quote); Ken Hofmeyer, "Growing Row Crops without Cultivation," *Wallaces Farmer*, 23 April 1966 (Engelkes quote).

40. Richard Krumme, Phil Jones, and Charles W. Sommers, "Your 9 Big Corn Decisions," *Successful Farming* (November 1970).

3—Fertilizer Gives the Land a Kick

1. *Wallaces Farmer* articles "Dollar Corn Makes Fertilizer Pay," 15 January 1944 (quote), and "Farmers Talk about Fertilizer Use," 20 January 1945.

2. Adams Papers; "Says Use More Fertilizer," *Wallaces Farmer*, 19 January 1946; Zenas H. Beers, "Development of the Fertilizer Industry in the Middle Western States," *Fertilizer Review* 26 (April–June 1951), 5; Bogue, *From Prairie to Corn Belt*, 285–86; "Fertilizers and Lime in the United States: Resources, Production, Marketing, and Use," in USDA *Miscellaneous Publication No. 586* (Washington, DC: GPO, May 1946), 26, 28, 85; "Fertilizers Move West," *Wallaces Farmer*, 6 September 1947 (quote); "More Farmers Use Fertilizer," *Wallaces Farmer*, 19 April 1952.

3. Willard W. Cochrane, *The Development of American Agriculture: A Historical Analysis* (Minneapolis: University of Minnesota Press, 1979), 127–28; "Consumption of Commercial Fertilizers, Primary Plant Nutrients, and Micronutrients," in USDA Crop Reporting Board, Statistical Reporting Service, *Statistical Bulletin No. 472* (Washington, DC: GPO, 1971).

4. The three numbers that describe "high analysis" fertilizer represent the percentages of nitrogen, phosphorous, and potassium per hundred pounds. Thus, 0-9-27 fertilizer was a mix of 0 percent nitrogen, 9 percent phosphorous, and 27 percent potassium, with the rest of the blend being a carrier. "Farmers Get Results," *Jefferson County Annual Report, 1958,* 11.

5. *Wallaces Farmer* articles "More Farmers Use Fertilizer," 19 April 1952, and "More Fertilizer Every Year," 7 August 1954; *Jefferson County Annual Report, 1957,* 6 (quote); Newt Hawkinson, "Starter Fertilizer," *Wallaces Farmer,* 20 April 1963.

6. Hawkinson, "Starter Fertilizer," *Wallaces Farmer,* 20 April 1963.

7. M. A. Anderson and E. L. Baum, "How Your Neighbors Use Fertilizer," *Iowa Farm Science* 11 (April 1957): 26–28; M. A. Anderson, L. E. Cairns, Earl O. Heady, and E. L. Baum, "An Appraisal of Factors Affecting the Acceptance and Use of Fertilizer in Iowa, 1953," *Special Report No. 16,* Agricultural Experiment Station, Iowa State College, Ames, June 1956.

8. Hudson, *Making the Corn Belt,* 171; Firman E. Bear, *Soils and Fertilizers,* 4th ed. (New York: John Wiley and Sons, 1951), xi; "What Big Yields Do to Soil Fertility," *Wallaces Farmer,* 2 November 1963; G. F. Sprague and J. C. Cunningham, "Growing the Bumper Corn Crop," in *A Century of Farming in Iowa, 1846–1946* (Ames: Iowa State College Press, 1946), 41.

9. James O. Bray and Patricia Watkins, "Technical Change in Corn Production in the United States, 1870–1960," *Journal of Farm Economics* 46 (November 1964): 762–63; "Are You Using Enough Fertilizer?" *Wallaces Farmer,* 3 February 1962.

10. *Wallaces Farmer* articles "Apply Fertilizer with Planter," 16 April 1955 (Byrnes, p. 40), and "Fertilizer Keeps Corn Yields High," 7 November 1953 (Griffieon, p. 12).

11. Al Bull, "Cut Down on Fertilizer?" *Wallaces Farmer,* 3 March 1956.

12. "Guide to Fertilizer Use," *Pamphlet 193,* Iowa State College Agricultural Extension Service, Ames, March 1953; *Wallaces Farmer* articles "Put Starter in the Right Place," 20 April 1957, "Early Boost from Starter," 5 April 1958 (Webster County, p. 36), "Use Both Starter and Plow-Down Fertilizer," 1 March 1961 (Fosseen, p. 91).

13. *Wallaces Farmer* articles "More Farmers Use Fertilizer," 19 April 1952, and Hawkinson, "Starter Fertilizer," 20 April 1963.

14. *Wallaces Farmer* articles Homer Hush, "Yellow Corn Short of Nitrogen," 17 June 1950 (journalist, p. 5), and "For More Corn Add Nitrogen," 21 May 1955 (Wilson, p. 29).

15. *Wallaces Farmer* articles "Plan to Side-Dress Your Corn?" 2 February 1952, Hush, "Yellow Corn Short of Nitrogen," 17 June 1950, and "Good Year to Side Dress Corn," 20 May 1960; Roswell Garst, "'Bob' Garst's Latest 'Crusade,'" *Successful Farming* (July 1962); Rex Gogerty, "Corn after Corn after Corn," *Farmer's Digest* 34 (February 1971): 20–22.

16. "Give Corn Fields Gas," *Wallaces Farmer,* 15 July 1950 (Carlson); Haight/Whittlesey Papers, State Historical Society of Iowa, Des Moines; *Wallaces Farmer* articles "Use Nitrogen for More Corn," 15 May 1954, and "Side Dress Nitrogen for Your Corn," 5 May 1962 (quote).

17. "Figure Plant Food in Fertilizer," *Wallaces Farmer,* 4 January 1947; diary entry, 30 November 1947, Ray L. Gribben Diary, State Historical Society of Iowa, Iowa City.

18. "What Time for Fertilizer?" *Wallaces Farmer,* 15 March 1952.

19. *Wallaces Farmer* articles, "Plow under Corn Fertilizer," 7 June 1953 (Patterson, p. 15), and "Boost Your Fertilizer Profits," 16 April 1955 (Krauter, p. 11).

20. *Wallaces Farmer* articles "Use Fertilizer This Fall," 17 September 1949, "Spread Fertilizer in the Fall?" 15 September 1951, and "Fertilize This Fall for 1953 Crop," 20 September 1952; Lloyd Dumenil, "It's Smart to Fall Fertilize," *Successful Farming* (October 1953); Lloyd Dumenil, H. R. Meldrum, and John Pesek, "When to Fertilize? Fall or Spring," *Iowa Farm Science* 9 (September 1954): 3–6.

21. *Wallaces Farmer* articles "More Fertilizer Applied in Fall," 16 January 1954, "Fall Application of Fertilizer," 20 August 1955 (Franzeen, p. 18), and Bob Dunaway, "Fall Soil Fertility Program," 28 October 1967 (quotes, p. 12).

22. Jim Johnson, interview by author, 21 June 2004, Eldora, Iowa, tape recording; "Apply Fertilizer in the Fall," *Wallaces Farmer,* 17 August 1957 (quotes).

23. *Wallaces Farmer* articles "More Fertilizer Applied in Fall," 16 January 1954, and "Iowa Soil Gets More Fertilizer," 17 January 1959.

24. "Apply Nitrogen in the Fall?" *Wallaces Farmer,* 1 October 1955.

25. *Wallaces Farmer* articles "When Should You Apply Your Fertilizer?" 6 October 1962 (Pesek, p. 15), and "Fall Fertilizer Application Can Boost 1965 Yields," 19 September 1964.

26. *Wallaces Farmer* articles Al Bull, "Apply Nitrogen in the Fall," 10 September 1966 (quotes), "Apply Nitrogen This Fall," 14 October 1967, and "Fall Nitrogen Program Still Safe, Practical," 11 October 1969; "Making the Most from Fall-Applied Anhydrous Ammonia for Corn," *Pm-334,* Iowa State University Cooperative Extension Service, Ames, August 1966 (Voes).

27. Norman West and Monte Sesker, "Special Plowing and Fertilizer Considerations for This Fall," *Wallaces Farmer,* 10 October 1970.

28. Donald R. Kaldor and Earl O. Heady, "An Exploratory Study of Expectations, Uncertainty and Farm Plans in Southern Iowa Agriculture," *Research Bulletin 408,* Agricultural Experiment Station, Iowa State College, Ames, April 1954, 877; "Fertilizer Use for Efficient Crop Production," *Pamphlet 227,* Cooperative Extension Service, Iowa State University, Ames, revised March 1962. *Wallaces Farmer* articles "Nitrogen Helps Second Year," 15 March 1952, "Fertilizer Helps in Second Year," 5 December 1953, and "Extra Profit from Fertilizer Carryover," 15 February 1964.

29. "Fertilizer Helps in Dry Years," *Wallaces Farmer,* 3 April 1954, p. 28.

30. *Second Biennial Report of Iowa Year Book of Agriculture* (Des Moines, State of Iowa, 1956); *Third Biennial Report of Iowa Year Book of Agriculture* (Des Moines, State of Iowa, 1958); *Wallaces Farmer* articles "Iowa Soil Gets More Fertilizer," 17 January 1959, and "Use Carryover to Build Up Fertility," 1 March 1958 (quote, p. 64); Carl T. and Bertha Peterson Farm Records, 1950, 1956.

31. Al Bull, "Fertilizer Keeps Corn Yields High," *Wallaces Farmer,* 7 November 1953.

32. "Fertilizer No Help to Beans," *Des Moines Register,* 7 May 1950; "Fertilizers for Field Crops," *Pamphlet 112,* Agricultural Extension Service, Iowa State College, Ames, revised January 1947, 8; "Guide to Higher Soybean Yields," *Pamphlet 202,* Agricultural Extension Service, Iowa State College, Ames, May 1953; "A Quick Guide for Higher Soybean Yields," *Pamphlet 290,* Cooperative Extension Service, Iowa State University, May 1962; "Cutting Costs in Today's Farming," *Pamphlet 222,* Agricultural Extension Service, Iowa State College, Ames, February 1956. *Wallaces Farmer* articles Dick Albrecht, "Growing Soybeans," 2 May 1959, "How They Get Top Soybean Yields," 2 May 1964 (Armann, p. 54), "Apply Fertilizer for Your Soybeans?" 4 May 1963, and "How to Shoot for Top Yields in 1965 Soybeans," 1 May 1965.

33. *Wallaces Farmer* articles "How They Get Top Soybean Yields," 2 May 1964 (Accola, p. 54), Al Bull, "Why Aren't Soybeans Producing Top Yields?" 14 May 1966, "Top Soybean Growers Interviewed," 13 May 1967 (survey), and Monte Sesker, "85 Bushel Beans!" 11 November 1967 (Harms).

34. *Wallaces Farmer* articles, Al Bull, "Get Higher Yields from Your Soybeans," 13 May 1967 (quotes), Monte Sesker, "New Approaches to Fertilizing Beans," 25 May 1968, and "How Soybeans Respond to Fertilizer," 10 February 1968 (quote, p. 24).

35. Hal Rothman, *Saving the Planet: The American Response to the Environment in the Twentieth Century* (Chicago: Ivan R. Dee, 2000), 113, 126; Samuel P. Hays, *Beauty, Health, and Permanence: Environmental Politics in the United States, 1955–1985* (Cambridge: Cambridge University Press, 1989), 162–64; N. William Hines, *Nor Any Drop to Drink: Public Regulation of Water Quality* (Iowa City: Agricultural Law Center, College of Law, University of Iowa, September 1967), 799–859.

36. Morris and Johnson, "Pollution Problems," 92–97.

37. *Wallaces Farmer* articles Seeley Lodwick, "Farmers Must Control Water Pollution," 27 July 1968, p. 23, and Al Bull, "Are Fertilizers Polluting Our Streams?" 23 May 1970 (quote, pp. 12–13); Ralph Sanders, "Pollution—Your Problem, Too," *Successful Farming, 1971 Crop Planning Issue* (November 1970).

38. C. Robert Taylor, "An Analysis of Nitrate Concentrations in Illinois Streams," *Illinois Agricultural Economics* 13 (January 1973): 12; *Tenth Biennial Report of Iowa Year Book of Agriculture, 1970–1971,* 84–86.

39. *Wallaces Farmer* articles "43 Pct. of Farmers Use Fertilizer," 1 October 1949, "Iowa Soil Gets More Fertilizer," 17 January 1959, "Did You Test Your Soil?" 18 May 1957, Al Bull, "Farmers Will Use More Fertilizer on Corn in '69," 8 March 1969, and "Corn Fertilizer Paid Off," 7 April 1956.

40. Peterson Papers.

41. *Wallaces Farmer* articles "Fertilizer Is Big Factor in High Corn Yields," 24 February 1968, Bull, "Farmers Will Use More Fertilizer on Corn in '69," 8 March 1969, "Fertilizer Use Has Doubled in 6 Years," 28 February 1970; Schipull Papers; "Fertilizer Made the Difference," *Wallaces Farmer* 18 January 1964 (quote, p. 43). "Profitable Corn Production," *Pamphlet 409,* Cooperative Extension Service, Iowa State University, Ames, January 1968. For expenses, see Jeffrey Finke, "Nitrogen Fertilizer: Price Levels and Sales in Illinois, 1945–1971," *Illinois Agricultural Economics* 13 (January 1973): 35–36. Fred R. Taylor, "North Dakota Agriculture since World War II," *North Dakota History* 34 (April 1967): 56; *Wallaces Farmer* articles "Corn Growing Costs Run High!" 19 April 1958, and Ken Hofmeyer, "What Does It Cost to Grow Good Corn?" 9 September 1967.

42. "Iowa Fertilizer Use Climbs," *Wallaces Farmer,* 27 February 1971.

4—Feeding Chemicals

1. Articles in *Successful Farming:* Truman Henley, "New Drug—Big Future" (March 1944), R. Allen Packer, "Penicillin Is Saving Cows" (November 1946), and C. E. Hughes, "Latest Roundup on Hormone Feeding" (October 1946); *Wallaces Farmer* articles "Hormones for Milk," 3 July 1948, and "Sex Hormone Makes Heifers Get Fat," 5 March 1949.

2. Terry G. Summons, "Animal Feed Additives, 1940–1966," *Agricultural History* 42 (October 1968): 305–13; Sidney W. Fox, "What Are Antibiotics?" *Iowa Farm Science* 7 (October 1952); J. S. Russell, "Credits ISC Discovery on Growth Drug," *Des Moines Sunday Register,* 14 May 1950; *Wallaces Farmer* articles Wally Inman, "Help Spring Pigs to Catch Up," 5 May 1951, and "Pigs to Market Weeks Sooner," 20 October 1951.

3. "Antibiotics," *Successful Farming* (March 1951); "What's Lowdown on Aureomycin?" *Wallaces Farmer,* 7 June 1950; Vernon Vine and Claude W. Gifford, "APF Gives Up Some Secrets," *Farm Journal* (June 1950); Damon Catron and Dean Wolf, "New Hope for 20 Million Runts," *Successful Farming* (March 1951); Wolf,

"Latest on Wonder Drugs in Feeds: What They Do for Hogs," *Successful Farming* (April 1951, quote, p. 37); Homer Hush, "Makes Hogs of Runts," *Wallaces Farmer,* 5 May 1951.

4. *Wallaces Farmer,* Pfizer advertisements, 18 October 1952, 20 June 1953.

5. Max Bartlett to Damon Catron, n.d., Damon Catron to Max Bartlett, 15 June 1950, Damon Catron Papers, Box 5, Folder 3, Special Collections, Parks Library, Iowa State University.

6. Aaron R. Bowman to Catron, 23 April, Catron to Bowman, 19 May 1951, Catron Papers, Box 7, Folder 1.

7. "Drugs Can't Whip Old Lots," *Wallaces Farmer,* 18 August 1951.

8. *Wallaces Farmer,* Homer Hush, "Shoot Antibiotic into Pigs," 21 June 1952, and "Inject Antibiotics into Pigs?" 7 March 1953 (quote, p. 74).

9. "Wean 'Em Early?" *Wallaces Farmer,* 5 March 1955; Vaughn Speer, Gordon Ashton, Francis Diaz, and Damon Catron, "New ISC Pre-Starter '75'," *Iowa Farm Science* 8 (April 1954).

10. From the Catron Papers, Jack D. Waite to Damon Catron, 20 April, Catron to Waite, 6 May 1954 (Box 15, Folder 1); Eugene Fitz to Catron, 27 March, Catron to Fitz, 8 April 1955 (Box 16a, Folder 1); Wilbur H. Frye to Catron, 29 December 1955, Catron to Frye, 6 January 1956 (Box 19, Folder 2); Don Buckley to Catron, 12 January, Catron to Buckley, 13 February 1956 (Box 19, Folder 1).

11. *Wallaces Farmer* articles "They're Weaning Earlier," 19 March 1955 (poll), "Wean Pigs at What Age?" 18 August 1956, and "When to Wean the Pigs," 7 September 1957 (quote, p. 6).

12. A. V. Nalbandov, "Hormones—What They Will and Won't Do for Livestock," *Successful Farming* (November 1952); Robert G. Rupp, "'Hurry-Up' Hormone for Feeder Cattle," *Farmer's Digest* 18 (June–July 1954).

13. Wise Burroughs, C. C. Culbertson, and William Zmolek, "Questions and Answers about Adding Stilbestrol to Feeds for Growing and Fattening Beef Cattle," *Pamphlet 215,* Agricultural Extension Service, Iowa State College, Ames, November 1954; Summons, "Animal Feed Additives," 310.

14. *Wallaces Farmer* articles "Feed Steers Gain-Booster," 19 February 1955, "Feed Stilbestrol," 15 October 1955, and "Stilbestrol—How's It Doing?" 5 November 1955 (quotes); Cliff Johnson interview.

15. William Zmolek and Wise Burroughs, "62 Questions about Stilbestrol Answered," *Bulletin P-133,* Agricultural and Home Economics Experiment Station, Cooperative Extension Service, Iowa State University of Science and Technology, Ames, July 1963.

16. Alan I Marcus, *Cancer from Beef: DES, Federal Food Regulation, and Consumer Confidence* (Baltimore: Johns Hopkins University Press, 1994), ch. 2.

17. "Drugs Won't Replace Good Livestock Management!" *Wallaces Farmer,* 2 May 1959.

18. "To Clamp Down on All Farm Chemicals," *Wallaces Farmer,* 5 December 1959 (quote, p. 8).

19. "Get Good Results from Antibiotics," *Wallaces Farmer,* 21 March 1959.

20. Finlay, "Hogs, Antibiotics."

21. Ron Lutz, "Groups Cite Opposition to FDA Drug Ban Talk," *Wallaces Farmer,* 22 June 1968 (quote, p. 5).

22. Ron Lutz, "Crackdown on Residues in Meats," *Wallaces Farmer,* 8 August 1970 (quote, p. 16).

23. Ibid., pp. 16–16a.

24. *Wallaces Farmer* articles, "New Stilbestrol Research," 15 March 1958, and Lutz, "Crackdown on Residues in Meats," 8 August 1970 (quote, p. 16).

25. "Behavior Studies Related to Pesticides," *Special Report No. 49*, Agricultural and Home Economics Experiment Station, Cooperative Extension Service, Iowa State University, December 1966.

26. Ralph Sanders, "Beef Management," *Successful Farming* (January 1971); "Drug Withdrawal Certificate Clarified," *Wallaces Farmer*, 26 June 1971.

27. "Take Animals off Drugs before Slaughter Time," *Wallaces Farmer*, 14 August 1971.

28. Wil Groves, "Needed: A Common Sense Approach to Feed Additive Use," *Wallaces Farmer*, 14 August 1971 (quotes, p. 18); Gary Wall, "Feed Additives . . . Drug Crisis in Agriculture," *Iowa Agriculturist* (Spring 1972, quote, p. 9).

29. Wil Groves and Bob Dunaway, "Stilbestrol Threatened by Illegal Residues" *Wallaces Farmer*, 27 November 1971 (quote, p. 15).

30. "Feeders Not Asked for Drug Statement," *Wallaces Farmer*, 8 April 1972.

31. Clark Mollenhoff, "FDA Halts Use of DES in Livestock Feed," *Des Moines Register*, 3 August 1972; George Anthan, "Farm State Congressmen Fight DES Ban," *Des Moines Register*, 4 August 1972.

32. Don Muhm, "Cattlemen Urge Allowing 'Realistic' DES Amounts," *Des Moines Register*, 14 July 1972; "Care Needed in Implanting DES," Iowa State University Information Service, "For Immediate Release," Wise Burroughs Papers, Special Collections, Parks Library, Iowa State University; Virgil Oakman and Jerald Heth, "Farmers: 'We'll Use DES'," *Des Moines Tribune*, 3 August 1972 (quote, p. 13).

33. Rodney J. Fee, "Will You Spend Five Minutes on This Form to Save Additives?" *Successful Farming* (August 1972); "Hogmen Could Lose Some Feed Additives," *Wallaces Farmer*, 23 September 1972.

34. "Farmers Find Drug Residues Are Costly," *Wallaces Farmer*, 8 April 1972.

5—Push-button Farming

1. Tony Basso, "Pushbuttons by '60?" *Wallaces Farmer*, 21 January 1950.

2. *Forty-First Annual Iowa Year Book of Agriculture, 1940*, 319; *Fifty-Second Annual Iowa Year Book of Agriculture, 1951*, 349; "Lighter Chores in '60," *Wallaces Farmer*, 4 February 1950.

3. Schlebecker, *Whereby We Thrive*, 254, 302–3; Reuben W. Hecht, "Labor Used for Livestock," in USDA Agricultural Research Service, *Statistical Bulletin No. 161* (Washington, DC: GPO, May 1955), 9; Bill Giese and Don Muhm, "Dairying without Carrying," *Successful Farming* (February 1951).

4. "How They Make Dairying Pay," *Iowa Farm and Home Register*, 6 March 1955.

5. "Cut Cost of Milk Production," *Wallaces Farmer*, 21 February 1953 (quote, p. 30); James R. Borcherding, "New Parlor Let Him Double Herd Size," *Successful Farming* (March 1959); "Milk More Cows with Less Work," *Wallaces Farmer*, 6 June 1959 (quote, p. 29).

6. *Wallaces Farmer* articles "Boys Like to Milk Cows," 17 March 1956 (quote, p. 18), "Want 'Push-Button' Dairy Barn?" 21 October 1950, "Iowa Dairyman Tries New Milking Parlor," 15 March 1958 (quote).

7. Dick Hanson, "Less Work, More Profits with Milk Tanks," *Successful Farming* (August 1952); W. H. M. Morris and Henry A. Homme, "What Is Bulk Milk Handling?" *Iowa Farm Science* 8 (June 1954).

8. *Fifty-Second Annual Iowa Year Book of Agriculture, 1951*, 41; Henry A. Homme, Eddie Easley, and John Shaul, "If You're Thinking of Going Grade A," *Iowa Farm Science* 7 (November 1952).

9. Hanson, "Less Work, More Profits with Milk Tanks," *Successful Farming* (August 1952) (quote, p. 80); *Wallaces Farmer* articles "Bulk Tanks Reduce Costs," 19 October 1957, "Which Bulk Tank for You?" 1 March 1958 (quote), and "Dairymen Find Bulk Tanks Pay," 21 February 1959.

10. "How to Handle a Bigger Dairy Herd," *Wallaces Farmer,* 2 August 1958.

11. *Wallaces Farmer* articles Dick Hagen, "Drylot Dairying," 20 May 1961, and Ken Hofmeyer, "Dairying Is Changing," 5 June 1965.

12. *Wallaces Farmer* articles "Bulk Tanks Reduce Costs," 19 October 1957, "Dairymen Find Bulk Tanks Pay," 21 February 1959, and "Milk More Cows with Less Work," 6 June 1959.

13. Newt Hawkinson, "Dairy Setup Designed for Efficiency," *Wallaces Farmer,* 8 February 1962.

14. *Wallaces Farmer* articles "Dairy Survey Information," 1 August 1959, and Jim Rutter, "Automatic Choring Equipment," 18 July 1959. There is some contradictory evidence on the numbers of farmers with pipeline milking systems. One article highlighting a survey of dairy farmers indicated that as many as 11 percent of farmers used pipeline milking systems, while an article from 18 July indicated that only 5 percent of Iowa farmers used pipeline systems. While it is impossible to gauge the accuracy of either survey, they do indicate that there were many people who continued to produce fluid milk in traditional structures.

15. *Wallaces Farmer* articles, Dave Bryant, "Dairy Herds Getting Bigger—Fewer," 15 October 1955 (quotes), and "Dairying Faces Changing Times," 17 June 1961 (quote, p. 11).

16. *Wallaces Farmer* articles, Bryant, "Dairy Herds Getting Bigger—Fewer," 15 October 1955 (quotes), and Ken Hofmeyer, "How Many Cows Should You Milk?" 5 October 1963 (quote, p. 59).

17. *Fifty-First Annual Iowa Year Book of Agriculture, 1950,* 570; *Fifth Biennial Report of Iowa Book of Agriculture, 1960–1961,* 351; Hagen, "Dairying Faces Changing Times," *Wallaces Farmer,* 17 June 1961.

18. Dick Hagen, "Goodby to Milk Cans," *Wallaces Farmer,* 21 January 1961. John T. Schlebecker noted that production and efficiency in American dairying increased 98 percent per man hour between 1940 and 1958, in *A History of American Dairying* (Chicago: Rand McNally, 1967), 40.

19. Thayer Cleaver and Robert G. Yeck, "Loose Housing for Dairy Cattle," in USDA *Information Bulletin No. 98* (Washington, DC: GPO, 1953), 2.

20. *Successful Farming* articles J. Clifford Grant, "Why I Built the Barn I Didn't Want" (May 1955), and Sherwood Searle, "I Switched to Loose Housing for $387" (April 1955).

21. "Saving Labor in Dairy Operations," n.d., Dale O. Hull Papers, Special Collections, Parks Library, Iowa State University; "Cut Dairy Chores," *Wallaces Farmer,* 1 November 1952 (Winkel, p. 8).

22. "Loose Housing Shows Promise," *Wallaces Farmer,* 1 March 1958.

23. "Loose Housing for Dairy Cattle?" *Wallaces Farmer,* 1 March 1953 (quote, p. 8); "Cut Dairy Chores," *Wallaces Farmer,* 1 November 1952.

24. "Free Stalls for Dairy," *Agricultural Engineers' Digest* (February 1963), Records of the Midwest Plan Service, Special Collections, Parks Library, Iowa State University; *Wallaces Farmer* articles "'Comfort' Stalls for Dairy Cows," 7 September 1963, and "Dairymen Like Free-Stall Housing," 10 February 1968 (quotes, p. 81).

25. Dick Hanson, "Work-Planned Cattle Feeding System," *Successful Farming* (October 1950).

26. R. N. Van Arsdall and Vernon Schneider, "Faster, Easier Feed Handling," *Successful Farming* (February 1954); Newt Hawkinson, "When Does It Pay to Mechanize?" *Wallaces Farmer,* 15 October 1960.

27. Cliff Johnson interview.

28. Ibid.; "Make Cattle Feeding Easier and Quicker," *Wallaces Farmer,* 15 February 1958 (quote, p. 62).

29. "Self Feeders Cut Choring Time," *Wallaces Farmer,* 7 July 1962.

30. *Wallaces Farmer* articles Dick Hagen, "Grain Banks Save You Labor," 3 January 1959 (quote, p. 12), and "Self Feeders Cut Choring Time," 7 July 1962 (quote, p. 30).

31. "Fifteen Minutes to Feed 135 Cattle," *Iowa Farm and Home Register,* 2 May 1954 (p. 5-H).

32. "Silo Unloaders," *Wallaces Farmer,* 21 November 1959.

33. Dick Hagen, "Mechanize Your Feedlot?" *Wallaces Farmer,* 17 October 1959; Pat Kellogg, "He Feeds 170 Head in 12 Minutes," *Successful Farming* (February 1959).

34. *Wallaces Farmer* articles Dick Hagen, "Mechanize Your Feedlot?" 18 February 1961 (quote, p. 14), "Saves Labor with Mechanized Feeding," 1 April 1961 (quote, p. 20), and "Let Augers Do the Job," 3 September 1960 (quote, p. 47); *Successful Farming Materials Handling, Third Edition,* circa 1960; Hull Papers, Box 4, Folder 1.

35. *Wallaces Farmer* articles "Feed Cattle with Less Work," 15 December 1956 (quote, p. 57), and "Here's a New Idea in Feeding," 4 April 1959.

36. Phil B. Jones, "He Built a '65 Feed Lot around a 1939 Barn," *Successful Farming* (September 1965).

37. Eugene S. Hahnel, Stilbosol advertisement, *Successful Farming* (December 1959); *Wallaces Farmer* articles "Augers Can Speed Choring Time," 3 March 1962 (Peet), "Let Augers Do the Job," 3 September 1960, Hagen, "Mechanize Your Feedlot?" 18 February 1961, "Silo Unloaders," 21 November 1959 (quotes, p. 14).

38. "They Mechanized Their Feedlot," *Wallaces Farmer,* 6 April 1963.

39. Wayne Messerly, "Automation Lets Him Feed More Cattle," *Wallaces Farmer,* 5 March 1966 (quote, p. 30).

40. "Piece by Piece They Built a Modern Feedlot," *Electricity on the Farm* (June 1967), Hull Papers, Box 3, Folder 31.

41. "Saves Labor with Mechanized Feeding," *Wallaces Farmer,* 1 April 1961; *Operation Feedbunk,* 8mm (quote) (Iowa State College Film Production Unit, 1959), Special Collections, Parks Library, Iowa State University.

42. "Outside Stuff," *Wallaces Farmer,* 4 March 1961 (quote, p. 66).

43. Newt Hawkinson, "Mechanized Feeding for Beef Cattle," *Wallaces Farmer,* 21 March 1964 (quote, p. 28).

44. "Cattle Feeding Methods Vary Widely in Iowa," *Wallaces Farmer,* 18 September 1965.

45. "Trend toward Larger Cattle Feedlots Continues," *Wallaces Farmer,* 28 March 1970.

46. Chris Mayda, "Pig Pens, Hog Houses, and Manure Pits: A Century of Change in Hog Production," *Material Culture* 36 (Spring 2004): 18–42.

47. "Confine 'Em or Turn 'Em Out?" *Wallaces Farmer,* 19 February 1955 (quote, p. 6).

48. "Which Way with Hogs?" *Wallaces Farmer,* 18 May 1957 (p. 16).

49. "Raise Hogs on Pasture or Concrete?" *Wallaces Farmer,* 19 April 1958.

50. "How to Make Hog Finishing Easy," *Successful Farming* (March 1959).

51. "New Concept in Hog Housing," Farm Business Report, *Wallaces Farmer,* 21 March 1964.

52. *Wallaces Farmer* articles Dick Hagen, "What Housing for Feeder Pigs?" 18 April 1959, Newt Hawkinson, "Confinement or Pasture for Hogs?" 7 April 1962, "Augers Move Both Feed and Manure," 21 July 1962, and "Timely Tips," 4 August 1962.

53. John Harvey, "3 Buildings, 1,200 Hogs—All Under Roof," *Successful Farming* (April 1965).

54. "Shelter for Beef Can Boost Gains," *Wallaces Farmer,* 1 November 1958 (quote, p. 22).

55. Chester Peterson, Jr., and William J. Fletcher, "Cattle Feeding Goes Inside," *Successful Farming* (September 1962); *Wallaces Farmer* articles Newt Hawkinson, "Outside Stuff," 7 July 1962 (quote, p. 44), and "Feeding Beef Cattle in Complete Confinement," 2 February 1963.

56. "Beef Confinement," *Successful Farming* (May 1970).

57. John Dorr, "Iowa's Largest Feedlot: Pampered Beef, Inc.," *Iowa Agriculturist* (Winter 1972); "Beef Confinement," *Successful Farming* (May 1970).

58. "Building Better Hogs," *Wallaces Farmer,* 1 September 1962.

59. *Wallaces Farmer* articles "Do You Grow Hogs in Confinement," 8 October 1966, Newt Hawkinson, "Confinement or Pasture for Hogs?" 7 April 1962, and Ken Hofmeyer, "Producers' Views on Hog Growing Systems," 6 February 1965 (quote, p. 56).

60. *Wallaces Farmer* articles "Do You Grow Hogs in Confinement," 8 October 1966 (quote, p. 71), and Hofmeyer, "Producers' Views on Hog Growing Systems," 6 February 1965 (quotes, p. 56).

61. *Wallaces Farmer* articles "Timely Tips," 19 December 1964 (quote, p. 9), and "Who's Raising All Our Hogs?" 20 February 1965.

62. Hofmeyer, "Producers' Views on Hog Growing Systems," *Wallaces Farmer,* 6 February 1965 (quote, p. 56); Carl Frederick, "How I Breed and Feed for No. 1 Hogs," *Successful Farming* (July 1965) (quote, p. 27).

63. "Hogs in Confinement Need Top Management," *Wallaces Farmer,* 25 January 1969 (quote, p. 75).

64. *Wallaces Farmer* articles "Farm Progress Show Confinement Beef Barn Pleases Owners," 14 March 1970 (quote, p. 124), and "Confinement Setup for Feeder Pigs," 27 January 1968.

65. Cooperative Extension Service, "Swine Equipment Plans" (Ames: Midwest Plan Service, Iowa State University, 1959). In 1965, livestock in Iowa created approximately 90–100 million pounds of manure, half of which was deposited in pastures while the other half was deposited in enclosures. Observers noted that the proportion of manure deposited in enclosures would increase rapidly with the rise of confinement feeding. George M. Browning, "Agricultural Pollution—Sources and Control," in *Water Pollution Control and Abatement,* ed. Ted L. Willrich and N. William Hines (Ames: Iowa State University Press, 1967), 159.

66. Dick Hagen, "Easy Manure Handling," *Wallaces Farmer,* 21 November 1959.

67. Newt Hawkinson, "What They're Learning about Manure Lagoons," *Wallaces Farmer,* 18 April 1964.

68. "New Approach to Manure Handling," *Wallaces Farmer,* 2 June 1962; T. L. Willrich and John Harvey, "Easier Hog-Manure Handling," *Successful Farming* (October 1962); Newt Hawkinson, "Manure Lagoons," *Wallaces Farmer,* 6 July 1963.

69. Page L. Bellinger, "Who Will Plan Your System?" *Successful Farming* (January 1962); "An Outline for Planning Your Cattle Feeding Facilities," *A.E. 954,* Iowa State University Cooperative Extension Service, Ames, January 1963, Hull Papers; "Dairy, 1969 Edition," *Midwest Plan Service Structures and Environment Handbook,* abstracted from "Dairy Equipment Plans and Housing Needs" MWPS-7, 1968, Records of the Midwest Plan Service; "Two-Year Study of Environmental Effects (Shelter and Paving) on Feedlot Steers," *A.S. Leaflet R52,* Iowa State University Cooperative Extension Service, Ames, October 1963, Hull Papers.

70. "Handling Swine Manure," *Agricultural Engineers' Digest* (Ames: Midwest Plan Service, 1969), 3, "Anerobic [*sic*] Manure Lagoons," *Agricultural Engineers' Digest* (Ames: Midwest Plan Service, 1969), and "Waste Disposal, 1969 Edition," *Midwest Plan Service Structures and Environment Handbook* (Ames: Midwest Plan Service, 1968), WD-14, all in Records of the Midwest Plan Service; Ted L. Willrich, "A Progress Report on Lagoons for Manure Disposal," n.d. (circa 1960–1961), Hull Papers.

71. Dale O. Hull to Rod Lorenzen, 14 June, 15 September (quote) 1966, Hull Papers, Box 2.

72. Ted L. Willrich to Rod Lorenzen, 27 September 1966 (quote), and "Statement for Group 21, Inc.—Re: Treatment and Disposal of Surface Runoff from Cattle Lots and Roads," Hull Papers, Box 2, Folder 22.

73. Al Bull, "Pollution Control for Cattle Feedlots," *Wallaces Farmer*, 11 May 1968 (quote, p. 18).

74. Ibid., and "Timely Tips," *Wallaces Farmer*, December 1970 (quote, p. 66).

75. Al Bull, "Water Pollution Control Hearings Aimed at Iowa's Cattle Feedlots," *Wallaces Farmer*, 13 April 1968; Tom Patrick, "Attack Proposed Feedlot Rules," *Des Moines Register*, 24 April 1968 (quote, p. 12).

76. Iowa State Department of Health, Water Pollution Control Commission, *Water Pollution Control Progress Report, 1968–1969* (Iowa Water Pollution Control Commission, State Department of Health, Des Moines), 12; Wil Groves, "Iowa's Feedlot Waste Disposal Law," *Wallaces Farmer*, 14 August 1971.

77. Iowa State Department of Health, *Water Pollution Control Progress Report, July 1, 1969–June 30, 1970*, 17, and *July 1, 1970–June 30, 1971*, 17.

78. *Wallaces Farmer* articles "Planning Feedlot Waste Disposal," 22 January 1972, and Wil Groves, "Step-by-Step Plan for Livestock Waste Control," 26 February 1972. The REAP program, formerly known as the Agriculture Conservation Program, provided up to $2,500 to farmers on a 50 percent cost-sharing basis to speed farmers' acceptance of the new regulations. *Wallaces Farmer* articles "Cost-Sharing Program Gets Pollution Control Emphasis," 13 March 1971, and Wil Groves, "Livestock Waste Control Systems," 22 January 1972.

79. "These Livestock Feeders Stop Pollution," *Wallaces Farmer*, 24 October 1970.

80. Ibid.

81. Stewart W. Melvin to Dewey Bondurant, 5 August 1970, Hull Papers, Box 3, Folder 30.

82. *Wallaces Farmer* articles Bob Dunaway, "What Cost to Stop Feedlot Pollution," 27 February 1971, and "These Livestock Feeders Stop Pollution," 24 October 1970 (quotes, p. 2).

83. "1970 Iowa Master Farmers," *Wallaces Farmer*, 24 January 1970.

6—Making Hay the Modern Way

1. Albert P. Brodell and Martin R. Cooper estimated that in 1944 approximately 75 percent of hay stored on farms in the United States was loose, long hay. "The Costs and Ways of Making Hay," in *Grass: The Yearbook of Agriculture, 1948* (Washington, DC: GPO, 1948), 173.

2. W. R. Humphries and R. B. Gray, "Partial History of Haying Equipment," in USDA *Information Series No. 74* (Washington, DC: GPO, revised October 1949), 55–56; O. F. Scholl, "The Twine Baler," *Agricultural Engineering* (November 1947): 501.

3. "Making Hay the Modern Way," *Wallaces Farmer*, 6 July 1946; Richard George Schmitt, Jr., "Economic Analysis of Haying Methods in Eastern Iowa" (Master's thesis, Iowa State College, 1947), 52–54; *Wallaces Farmer* articles "How to Handle Your Hay," 2 June 1951, and "Half of Farmers Bale All Hay," 19 July 1952.

4. "That Hired Man," *Wallaces Farmer*, 15 January 1944; Marcus and Segal, *Technology in America*, 301; J. B. Davidson and D. K. Struthers, "Haying with Less Help," *Farm Science Reporter* 4 (April 1943), 3–6; "How to Handle Your Hay," *Wallaces Farmer*, 2 June 1951 (quote, p. 14).

5. John Deere advertisement, *Wallaces Farmer*, 15 February 1958.

6. *Wallaces Farmer* articles "Workday Pointers," 20 June 1953, and "Two Can Make Hay," 4 September 1948.

7. *Wallaces Farmer* articles "Cultivate Corn or Make Hay?" 4 June 1955, and "Can You Guess the Weather?" 20 July 1957 (quote).

8. Kenneth K. Barnes and Robert C. Fincham, "Yes, Hay Crushers Can Cut Field Risks," *Iowa Farm Science* 12 (May 1958). *Wallaces Farmer* articles Dick Seim, "Early-Cut

Hay Is Best," 16 May 1959 (quote, p. 14), "Conditioned Hay Dries Faster," 21 May 1960, and "For Faster Haying and Better Hay," 23 May 1970 (quote, p. 49).

9. *Midwest Farm Handbook, Fifth Edition* (Ames: Iowa State University Press, 1960), 613.

10. "Forage Harvesting Aimed at Quality," *Wallaces Farmer*, 2 May 1959.

11. *Wallaces Farmer* articles "Take Sweat and Risk from Hay-Making," 7 August 1948, Wendell Clampitt, "Save Sweat, Dust in Haymaking," 17 June 1950 (quote), Jim Rutter, "Grass Silage or Hay?" 3 May 1958.

12. *Wallaces Farmer* articles Jim Rutter, "Will You Chop or Bale?" 7 June 1958 (quote, p. 38), and Rutter, "Grass Silage or Hay?" 3 May 1958 (quote, p. 11).

13. Doyle Brubaker, interview by author, 7 September 2001, Newton, Iowa, tape recording. In *Wallaces Farmer*, Allis Chalmers Roto-Baler advertisement, 2 April 1955, "Stacking Trims Haying Costs," 2 June 1962, and "Which Haying Method for Beef Cattle?" 2 June 1962.

14. *Wallaces Farmer* articles "New Hay Machines Save Labor," 16 August 1949, and "What Cost for Making Hay?" 6 June 1955; Adams Papers, 12 June 1952.

15. J. Sanford Rikoon demonstrated how new threshing machines actually complemented and even strengthened work exchange patterns in the nineteenth-century Midwest. Mary Neth depicted the importance of work exchanges for minimizing cash expenses in the years up to 1940. See Rikoon, *Threshing in the Midwest: A Study of Traditional Culture and Technological Change, 1820–1940* (Bloomington and Indianapolis: Indiana University Press, 1988), and Neth, *Preserving the Family Farm*. "Make Hay Together," *Wallaces Farmer*, 16 June 1945 (quote, p. 14).

16. *Wallaces Farmer* articles "What's Best Way to Get Quality Hay?" 7 June 1947, and "Exchanging Work with Neighbors," 6 July 1957.

17. Havran interview.

18. Schmitt, "Haying Methods in Eastern Iowa," 56; Melvin Hansen, interview by author, 9 June 2003, Madrid, Iowa, tape recording; "Own Machines in Partnership," *Wallaces Farmer*, 1 May 1954 (quote).

19. Schmitt, "Haying Methods in Eastern Iowa," 48; "Can You Guess the Weather?" *Wallaces Farmer*, 20 July 1957; Peterson Papers.

20. "Exchanging Work with Neighbors," *Wallaces Farmer*, 6 July 1957; Peterson Papers.

21. Jim Rutter, "Will You Chop or Bale?" *Wallaces Farmer*, 7 June 1958.

22. *Wallaces Farmer* articles "How Much to Hire a Combine?" 19 June 1948, "How to Handle Your Hay," 2 June 1951 (quote, p. 14); Peterson Papers; Richard Hagen, "How Much Money for Machinery?" *Wallaces Farmer*, 2 June 1962 (quote, p. 38).

23. *Wallaces Farmer* articles "Forage Harvesting Aimed at Quality," 2 May 1959, and Al Bull, "Make Hay or Silage?" 15 May 1954.

24. *Wallaces Farmer* articles "Want to Stretch Your Pasture?" 15 June 1954, Dave Bryant, "Stretch Pasture with 'Zero' Grazing?" 7 May 1955 (quotes, p. 11), and "Carry Pasture to Your Cattle," 6 May 1956; Vernon Schneider, "Green-Lot Feeding," *Successful Farming* (March 1955).

25. Hagen, "Drylot Dairying," *Wallaces Farmer*, 20 May 1961 (quote, p. 11).

26. Ray Franklin, "Cheaper Milk from Forage," *Wallaces Farmer*, 21 September 1957 (quote, p. 11).

27. "Old Barns," *Wallaces Farmer*, 21 July 1951.

28. *Wallaces Farmer* articles Jim Rutter, "Will You Chop or Bale?" 7 June 1953, and "Easy Ways to Move Hay," 21 June 1958. Gerald L. Kline, "Harvesting Hay with the Automatic Field Baler" (Master's thesis, Iowa State College, 1946), 78.

29. R. B. Gray, "Equipment for Making Hay," in *Grass*, 169; Dave Bryant, "Do You Need a Hay Dryer?" *Wallaces Farmer*, 7 June 1952 (quote, p. 14).

30. *Wallaces Farmer* articles Bryant, "Do You Need a Hay Dryer?" 7 June 1952 (Parrett), Keith Remy, "How to Get the Most from Your Hay," 21 May 1960 (Mather, p. 12), and Dick Seim, "Early-Cut Hay Is Best," 16 May 1959.

31. Gene C. Shove, Kenneth K. Barnes, and Hobart Beresford, "A Self-Feeding Hay-Storage Structure," *Iowa Farm Science* 8 (June 1954); "Cut Hay Handling with Hay-keeper," *Wallaces Farmer,* 15 August 1959 (quote, p. 68); Glenn Cunningham, "They Gross $50,000 on 240 Acres," *Iowa Farm and Home Register,* 6 March 1955.

32. *Wallaces Farmer* articles "A Silo in a Hurry," 15 July 1950, and "Temporary Silo for You?" 4 September 1954.

33. "Builds Low-Cost Shed for Storing Hay," *Wallaces Farmer,* 21 July 1962.

34. *Fifth Biennial Report of Iowa Book of Agriculture, 1960–1961* (Des Moines: State of Iowa, 1962), 353; *Iowa Book of Agriculture, 1970–1971,* 368.

35. "Giant Hay Bale Will Mechanize Handling," *Wallaces Farmer,* 26 August 1967.

7—From Threshing Machine to Combine

1. R. Douglas Hurt, *American Farm Tools from Hand Power to Steam Power* (Manhattan, KS: Sunflower University Press, 1986), 77–80; Graeme Quick and Wesley Buchele, *The Grain Harvesters* (St. Joseph, MI: American Society of Agricultural Engineers, 1978), 85–90.

2. *Farm Implement News* articles "McCormick Deering Take-Off Combine," 27 March 1930, "John Deere No. 5 Combine," 29 May 1930, "The Fleming-Hall Baby Combine," 7 August 1930, and "Allis-Chalmers Announces 'Baby' Combine," 20 November 1930; Charles H. Wendel, *The Allis-Chalmers Story* (Osceola, WI: Crestline Publishing, 1993), 66.

3. W. M. Hurst and W. R. Humphries, "Performance Studies of Small Combines," *Agricultural Engineering* 17 (June 1936): 249–50; W. M. Hurst, "The Field for the Small Combined Harvester-Thresher" *Agricultural Engineering* 16 (June 1935): 221–22, and discussion by F. N. G. Kranick, 222–23.

4. Brubaker interview (quote); "Combines Gain on Binders," *Wallaces Farmer,* 10 August 1940.

5. Brubaker interview.

6. Quick and Buchele, *Grain Harvesters,* 172–73; Keith Robinson, interview by author, 1 October 1992, Atlantic, Iowa, tape recording; Lennis J. Holm, "From Family Farm to Agribusiness" (Honors thesis, University of Iowa, May 1967), 115; Darlene and Elmer Meyer, letter to author, 7 September 2001; "John Deere No. 12-A Combine Paves the Way to Bigger Profits," John Deere Sales Brochure, A-534-50-8, 1950.

7. "Own or Hire the Combine," *Wallaces Farmer,* 21 June 1952.

8. Ibid.; "Should You Own a Combine?" *Wallaces Farmer,* 5 June 1954; "Are You Losing Money on Machinery?" *Successful Farming* (May 1950).

9. Schipull Papers.

10. The expenses for combining, binding, and threshing do not include fuel, which Schipull did not itemize by type of fieldwork. If figures for fuel were available, the cost of combining would be even more favorable compared to that of binding and threshing, since the latter required a trip across the field for harvesting, picking up bundles, and at least two days of threshing. Schipull Papers.

11. Schipull Papers, 7, 8 August 1947; Ludwig Papers, 27 July 1948.

12. "Has Threshing Lost Its Glamour?" *Wallaces Farmer,* 21 July 1945.

13. Johnnie Westphalen, interview by author, 2 October 1992, Atlantic, Iowa, tape recording.

14. Rikoon argued that farmers continued to thresh after World War II because they "weighed the financial cost of innovation and concluded that older practices still

provided the most efficient way to complete the grain harvest" (*Threshing in the Midwest*, 153). Holm, "From Family Farm to Agribusiness," 115; Robinson interview.

15. *Wallaces Farmer* article "Combines Gain on Binders," 10 August 1940 (quotes, p. 12).

16. "No More Threshing Ring Dinners?" *Wallaces Farmer*, 27 June 1942.

17. "Let Grain Get Ripe, Farmers Say," *Wallaces Farmer*, 6 July 1946.

18. *Wallaces Farmer* articles "Windrowing Helps Grain Quality," 18 June 1949, "Windrowed Grain Keeps Better," 1 July 1950, "Need Straw Next Winter?" 21 July 1951, and "Oat Harvest Method Varies over State," 20 June 1959; "Conditioning Grain for Harvesting with the Combine," Hull Papers, Box 9, Folder 9. In 1960 the author of a brief article titled "Harvesting Oats" asked whether farmers should combine standing oats or from a windrow without offering any indication of a preferred method. *Wallaces Farmer*, 2 July 1960.

19. "No More Threshing Ring Dinners?" *Wallaces Farmer*, 27 June 1942 (quote); Hull, "Conditioning Grain for Harvesting with the Combine," Hull Papers, Box 9, Folder 9; Charles and Minnie Havran Farm Records, Living History Farms, Urbandale, Iowa; Brubaker interview. A businessman from Bettendorf, Iowa, offered farmers free instructions for converting binders to windrowers beginning in 1940. See George Innes, "Old Binders Make Good Windrowers," *Farm Implement News*, 2 May 1940, 33.

20. Gribben Diary, 7/29, 7/25, 7/26.

21. Robinson interview (quote); Edward O. Moe and Carl G. Taylor, "Culture of a Contemporary Rural Community, Irwin, Iowa," *Rural Life Studies 5* (Washington, DC: GPO, December 1942), 39, 78.

22. "Getting onto Combines," *Wallaces Farmer*, 14 June 1941 (quote); Gribben Diary, 7/28, 7/29; R. D. Anderson, interview by author, 7 December 2001, Des Moines, Iowa, tape recording; "Need Straw Next Winter?" *Wallaces Farmer*, 21 July 1951.

23. Mary Neth stressed the importance of threshing as a farm survival strategy that allowed family farms to minimize expenses by exchanging labor within their communities. According to Neth, the breakup of threshing rings was a loss of community sharing that men and especially women mourned. I do not deny that farm family members valued aspects of threshing, but it is likely that they welcomed the new work more than they mourned the passing of the old. See *Preserving the Family Farm*, ch. 6. Elmer and Darlene Meyer, interview by author, 14 August 2001, Bridgewater, Iowa, tape recording; Robinson interview; Anderson interview.

24. "Streamlined Threshing Meals," *Wallaces Farmer*, 27 June 1940.

25. "Feeding the Threshers," *Wallaces Farmer*, 27 June 1942.

26. "Extra Men Coming for Dinner," *Wallaces Farmer*, 15 July 1950; Meyers interview.

27. Jellison, *Entitled to Power*, 140; "Oat Harvest," *Wallaces Farmer*, 7 August 1945; Meyers interview; Deborah Fink, *Open Country: Rural Women, Tradition, and Change* (Albany: State University of New York Press, 1986), ch. 7.

28. *The Sheppards Take a Vacation*, produced by Ray-Bell Films for John Deere Company, directed by Reid H. Ray, circa 1940.

29. "How to Take Eight on Your Vacation," *Wallaces Farmer*, 15 July 1950 (quote); Anderson interview.

30. *Wallaces Farmer* articles "Power Farming Changes Leases," 5 October 1946, "Who Pays: Tenant or Landlord?" 3 July 1948, and "Who Pays for Combining?" 1 July 1950.

31. Earle D. Ross, *Iowa Agriculture* (Iowa City: State Historical Society of Iowa Press, 1951), 180; Wilcox, *Farmer in the Second World War*, 299; "We Must Have More Soybeans and Flax!" *Pamphlet 26*, Agricultural Extension Service, Iowa State College, Ames, March 1942; "Is the Soybean Boom Over?" *Wallaces Farmer*, 20 October 1945; *U.S. Census of Agriculture, 1940*, vol. 1, pt. 2, p. 121; "Combines on Beans," *Wallaces Farmer*, 15 September 1942; *Second Biennial Report of Iowa Book of Agriculture, 1954–1955*, 406; *Fifth Biennial Report of Iowa Book of Agriculture, 1960–1961*, 328.

32. "How to Get Those Soybeans In," *Wallaces Farmer,* 18 September 1943; Rikoon included a skillful discussion of the importance of combines in the transition to soybeans in *Threshing in the Midwest,* ch. 7. Ross, *Iowa Agriculture,* 180; Lillian Church, "Partial History of the Development of Grain Harvesting Equipment," *Information Series No. 72* (Washington, DC.: GPO, revised October 1947), 51; Schipull Papers; Asa T. Meelchryst Papers, State Historical Society of Iowa, Des Moines, Iowa.

33. "Thresh Beans with Care," *Wallaces Farmer,* 17 October 1942.

34. *Wallaces Farmer* articles "Does Combine Save Beans?" 16 September 1944, "Lose One-Tenth of Bean Crop?" 15 September 1951 (quote), and "How to Save More Beans," 19 September 1954.

35. John Deere advertisement, *Farm Implement News,* 25 November 1950.

36. *First Biennial Report of Iowa Book of Agriculture, 1952–1953,* 475; *Fourth Biennial Report of Iowa Book of Agriculture, 1958–1959,* 329.

37. "Scanning the Self-Propelled Combines," *Implement and Tractor,* 21 June 1965, 27–28.

38. John Deere Combine advertisement, *Wallaces Farmer,* 17 May 1941; "Need Straw Next Winter?" *Wallaces Farmer,* 21 July 1951.

8—From Corn Picker and Crib to Combine and Bin

1. Carl Hamilton, *In No Time At All* (Ames: Iowa State University Press, 1974), 78–84.

2. Ibid., 82–83. *Wallaces Farmer* articles Beth Wilcoxson, "We'll Take Machines," 20 October 1956 (quote, p. 65), and "Thru Picking by Thanksgiving?" 16 November 1946.

3. Thomas Burnell Colbert, "Iowa Farmers and Mechanical Corn Pickers, 1900–1952," *Agricultural History* 74 (Spring 2000), 531; M. N. Beeler, "What Users Say about Their Corn Pickers," *Farm Implement News,* 21 September 1939 (quote, p. 32); Sid Dix, "The Arrival of the Corn Picker," *Farm Implement News,* 7 January 1943 (cartoon).

4. Gribben Diary, 3, 4, 6 November 1948.

5. Ibid., 10–14, 16, 24–26, 28–29 November 1948, 2, 3, December 1948.

6. Ludwig Papers.

7. Annual corn yields are included in the annual and later biennial editions of the *Iowa Year Book of Agriculture* and are also accessible on the Web site www.nass.usda.gov/ia/historic/crn1866.txt.

8. *Wallaces Farmer* articles "Got Enough Cribs?" 4 September 1948, "Corn Floods Iowa Farms," 20 November 1948, and "Need Another Crib for Extra Corn?" 18 October 1952. *Forty-Ninth Annual Iowa Year Book of Agriculture, 1948,* 10.

9. "How Many Buildings Are Modern?" *Wallaces Farmer,* 7 August 1948.

10. Al Bull, "What Kind of Crib Do You Need?" *Wallaces Farmer,* 6 September 1952 (quote, p. 8). For examples of the boom in corncrib construction and the types of buildings farmers erected, see the following articles in *Wallaces Farmer*: "You Can Still Plant Early Potatoes," 3 April 1948, "How Many Buildings Are Modern?" 7 August 1948, "Got Enough Cribs?" 4 September 1948, "Can Seal Corn in Round Cribs," 2 October 1948, Bull, "What Kind of Crib Do You Need?" 6 September 1952, "Where to Put Corn?" 15 August 1953, "When You Build Corn Storage," 17 July 1954 (quotes); Melvin Laughlin Papers, State Historical Society of Iowa, Iowa City; Schipull Papers.

11. For Quonset buildings see "Forced Air Drying for Cribs?" 16 August 1952, and Quonset advertisement, 19 September 1953; for the MWPS structure see "Crib Built for Drying Corn," 19 September 1953, and "When You Build Corn Storage," 17 July 1954, all in *Wallaces Farmer.*

12. "Need Another Crib for Extra Corn?" *Wallaces Farmer,* 18 October 1952; Laughlin Papers, 20 September.

13. *Wallaces Farmer* articles "Need Another Crib for Extra Corn?" 18 October

1952; "Ear Corn . . . Where You Want It," 4 September 1954.

14. "When You Build Corn Storage," *Wallaces Farmer,* 17 July 1954.

15. *Into Tomorrow,* produced by the Calvin Company for Massey-Harris Company, directed by Reese Wade, 1946.

16. "What They Say on Picker-Shellers," *Wallaces Farmer,* 5 September 1953 (quote, p. 24); Floyd L. Herum and Kenneth K. Barnes, "What's the Best Way to Harvest Corn?" *Iowa Farm Science* 9 (July 1954) (quote); Sherwood Searle and Vernon Schneider, "Here's Why We Like Field Shelling," *Successful Farming* (October 1953) (quote, p. 46). See also C. H. Van Vlack and Hobart Beresford, "The Picker-Sheller: Its Advantages and Disadvantages," *Iowa Farm Science* 8 (July 1953).

17. *Wallaces Farmer* articles "Corn Storage Plans," 6 September 1958 (quote, p. 28), and A. W. Ranniger, "Use Picker-Sheller and Dryer?" 17 September 1955 (quotes).

18. "How Will You Harvest Corn?" *Wallaces Farmer,* 15 September 1962.

19. The following articles from *Agricultural Engineering* describe the technical development of the corn head: L. W. Hurlbut, "More Efficient Corn Harvesting," "Laboratory Studies of Corn Combining," C. S. Morrison, "Attachments for Combining Corn," all in *Agricultural Engineering* 36 (December 1955). John Deere 45 Combine and Corn Attachment advertisement, *Wallaces Farmer,* 4 August 1956.

20. *Wallaces Farmer* articles "Pick Corn When You're Ready," 15 August 1959 (quote, p. 13), Jim Rutter, "Field Shell Your Corn?" 5 September 1959, "A Sign of the Times—Combining Corn," 3 October 1964 (quote, p. 45a), and Norman West, "Which Way to Harvest Corn?" 14 August 1971 (quote, p. 97).

21. *Wallaces Farmer* articles, Rutter, "Field Shell Your Corn?" 5 September 1959, "Outside Stuff," 17 October 1959 (quote), and "Trend Is toward More Field Shelling," 28 June 1969 (quote).

22. "Trend Is toward More Field Shelling," *Wallaces Farmer,* 28 June 1969; *Biennial Report of the Iowa Department of Agriculture, 1971–1972,* 38.

23. Ludwig Papers.

24. Ibid.; Monte Sesker, "What It Costs to Own a Combine," *Wallaces Farmer,* 14 September 1968.

25. *Wallaces Farmer* articles "When You Build Corn Storage," 17 July 1954, Dick Hagen, "Remodeled Cribs Can Double Your Storage," 3 October 1959 (quote, pp. 54–55), "Twice as Much Corn in the Crib," 1 October 1960 (quote, p. 47); Marsha and Ken Smalley, interview by author, 21 June 2004, Iowa City, Iowa, tape recording.

26. "Remodelling Corn Cribs for Small Grain Storage," *Agricultural Engineers' Digest, AED-12,* n.d., Records of the Midwest Plan Service.

27. *Wallaces Farmer* articles "Converting Cribs to Shelled Corn Storage," 2 September 1961 (quote, p. 45), "Store Shelled Corn in Remodeled Cribs," 2 August 1958 (quote, pp. 42–43), "Field Shell: Pick Corn When You're Ready," 15 August 1959 (quote, p. 13), and "Convert Corn Crib to Shelled Corn Bin?" 17 August 1963; Frank C. Beeson, "Lined Cribs Make Shelled-Corn Storage," *Successful Farming* (October 1959).

28. *Wallaces Farmer* articles A. W. Ranniger, "Use Picker-Sheller and Dryer?" 17 September 1955 (quote), "From Cow Barn to Corn Bin," 21 October 1961 (quote, p. 44), and "Converts Dairy Barn to Corn Storage," 6 February 1965 (quote, p. 38).

29. Behlen Manufacturing Company advertisement, *Wallaces Farmer,* 21 July 1956.

30. *Wallaces Farmer* articles "Corn Storage Plans," 6 September 1958, and "Timely Tips," 3 June 1961 (quote).

31. Newt Hawkinson, "It Pays to Plan before You Build Grain Storage," *Wallaces Farmer,* 19 August 1961 (quote, p. 13).

32. "Corn Storage Plans," *Wallaces Farmer,* 6 September 1958.

33. "Put Dry Corn in Your Cribs," *Wallaces Farmer,* 20 September 1958.

34. *Wallaces Farmer* articles "Engineers List Ways to Save Wet Corn," 2 August 1947, and "Fix Cribs to Dry Corn," 15 September 1945.

35. "Wet Corn Stopped Husking Rush," *Wallaces Farmer,* 1 December 1945; "Drying Ear Corn by Mechanical Ventilation," in USDA *Miscellaneous Publication No. 919* (Washington, DC: GPO, 1963), 3; "Looks like Moldy Corn for Hogs," *Wallaces Farmer,* 6 April 1946 (quote).

36. *Wallaces Farmer* articles "This Is the Year for Dryers," 7 October 1950, and "What's Ahead in Corn Storage?" 2 August 1952; *Fiftieth Annual Iowa Year Book of Agriculture,* 15.

37. "Tired of Docks on Corn?" *Wallaces Farmer,* 15 September 1951.

38. *Wallaces Farmer* articles Dave Bryant, "Harvest Corn Early!" 4 September 1954 (quote), "Use Picker-Sheller and Dryer?" 17 September 1955, "Put Dry Corn in Your Cribs," 20 September 1958 (quote), "Stop Wet Corn Spoilage Losses," 15 March 1958, and "Emergency Corn Drying," 5 April 1958. *Farmers' Manual of Crop Drying* (Agricultural Development Division of the Lennox Furnace Company, n.d.), Hull Papers.

39. "Storage of Small Grains and Shelled Corn on the Farm," in USDA *Farmers' Bulletin No. 2009* (Washington, DC: GPO, 1949); *Wallaces Farmer* articles "Put Dry Corn in Your Cribs," 20 September 1958 (quote), and "This Corn Crop Is a Problem!" 21 November 1959 (quote, p. 58).

40. *Wallaces Farmer* articles "Timely Tips," 22 August 1970, Frank Holdmeyer, "Bin Dryers . . . Portable Batch . . . Continuous Flow," 12 August 1971; "Boosting Profits with High-Moisture Corn," *Pioneer Corn Service Bulletin* (n.d., circa 1965), 5, Hull Papers; Norman West, "Match Your Dryer System to Harvest Rate," *Wallaces Farmer,* 28 August 1971.

41. Samuel R. Aldrich and Earl R. Leng, *Modern Corn Production* (Cincinnati, OH: Farm Quarterly, 1965), 280–83.

42. "Boosting Profits with High-Moisture Corn," 3, Hull Papers.

43. Monte Sesker, "Tips for Better Dryer Operation," *Wallaces Farmer,* 28 September 1968. A good example of the growing sophistication of grain storage and drying can be seen in *Planning Grain-Feed Handling for Livestock and Cash-Grain Farms, MWPS-13* (Ames: Midwest Plan Service and Iowa State University, 1968), Records of the Midwest Plan Service.

44. Sukup advertisement, *Farm Industry News* (February 2003); "Stirring Devices Will Speed Grain Drying," *Wallaces Farmer,* 23 September 1967.

45. Ludwig Papers; Ken Hofmeyer, "What Does It Cost to Grow Good Corn?" *Wallaces Farmer,* 23 September 1967 (quote).

46. *Biennial Reports of Iowa Year Book of Agriculture, 1964–1965, 1966–1967, 1968–1969, 1970–1971;* Ralph Sanders and Richard Krumme, "How These Farmers Harvest and Sell Corn," *Successful Farming, Harvesting Issue, 1970* (quote).

47. William Mason Family Farm Records (including papers of Foster and Madeline Mason), State Historical Society of Iowa, Des Moines.

48. *Wallaces Farmer* articles "Many Will Get New Cribs, Bins," 1 August 1953, and "Need More Corn Storage," 6 August 1955.

49. "A Federal Loan Helped Build This," *Iowa Farm and Home Register,* 2 May 1954; *Wallaces Farmer* articles "To Make Loans on Grain Dryers," 5 November 1949, "Government Grain Bins for Sale," 6 February 1965, "Loans for Bins and Dryers," 26 August 1967, and "Low Cost Loans for Grain Storage," 28 July 1968.

50. In *Wallaces Farmer,* "More Farmers Use Bins to Dry Corn," 7 September 1963, and Behlen Manufacturing Company advertisements, 21 July 1956, 26 April 1969.

51. Aldrich and Leng, *Modern Corn Production,* 280; Ludwig Papers; Newt Hawkinson, "Which Drying System for Shelled Corn?" *Wallaces Farmer,* 15 August 1964 (quote, p. 14).

52. "Harvest Show at History Farms," *Wallaces Farmer,* 24 October 1970; *Corn Harvest Festival,* 8mm, produced by the Living History Farms Foundation and the Iowa State University Film Production Unit, 1971, Special Collections, Parks Library, Iowa State University.

Conclusion

1. Jim Johnson interview (quotes); Donald R. Murphy, "It Ain't like the Good Ole Days!" *Wallaces Farmer,* 10 February 1968.

2. Tony Leys, "Scientists Suspect Pesticides May Play Role in Parkinson's," and "Study Connects Banned Pesticides to Brain Ailments," both in *Des Moines Sunday Register,* 15 May 2005.

3. *Des Moines Register* articles by Perry Beeman, "Many Cattle Feedlots Lack Needed Manure Controls," 4 April 2006, and "Panel OKs New Rules to Clean Up Iowa's Rivers," 18 January 2006; personal communication.

4. For graphic representations of the postwar cost-price squeeze, see Milton C. Hallberg, *Economic Trends in U.S. Agriculture and Food Systems since World War II* (Ames: Iowa State University Press, 2001), 41–42; Rasmussen, "Technological Change," 588.

5. Rasmussen, "Technological Change."

6. For a concise overview of the environmental movement in the United States, see Rothman, *Saving the Planet;* Fred Bailey, Jr., "From DES to Fertilizer—We're in a Fight for Survival," *Successful Farming* (May 1972).

7. "Big Rise in Farm Costs Shown in USDA Study," *Wallaces Farmer,* 26 February 1972.

8. Dick Hanson, "It Costs Less to Farm Big," *Successful Farming* (March 1970).

9. This trend toward part-ownership is discussed more fully in John Fraser Hart, "Part-Ownership and Farm Enlargement in the Midwest," *Annals of the Association of American Geographers* 81 (March 1991): 67–70.

10. Al Bull, "What Kind of Iowa Do We Want?" *Wallaces Farmer,* 13 November 1971. In a study of Iowa farmers from the mid-1970s, a rural sociologist found that approximately 40 percent of farmers were satisfied with current government regulation of agricultural chemicals and pollution-control measures, while 10–14 percent of farmers believed there was too little regulation. The poll addressed only the issue of regulation, however, not the technology that helped create the sense among rural and urban people that regulation was necessary. Farmers accepted that chemicals and machines were essential to farming. They did not question the value of new technology or believe it was bad; it was only the potential abuses and abusers of technology that were to be regulated. Eric O. Hoiberg and Wallace Huffman, "Profile of Iowa Farms and Farm Families, 1976," *Bulletin P-141,* Iowa Agriculture and Home Economics Experiment Station and Cooperative Extension Service, Iowa State University, Ames, April 1978, 14–15.

11. Denise O'Brien, personal communication.

Bibliography

Primary Sources

Archival Collections

LIVING HISTORY FARMS, URBANDALE, IOWA

Charles and Minnie Havran Farm Records.

SPECIAL COLLECTIONS, PARKS LIBRARY,
IOWA STATE UNIVERSITY

William M. Adams Papers.
Annual Narrative Reports of County Extension Agents. 1945–1968.
Annual Report for Entomology and Wildlife, Cooperative Extension Service. 1945–1967.
Wise Burroughs Papers.
Damon Catron Papers.
Corn Harvest Festival. 8mm. Living History Farms Foundation and the Film Production Unit, Iowa State University, 1971.
Dale O. Hull Papers.
Joseph Ludwig Papers.
Midwest Plan Service Records.
Operation Feedbunk. 8mm Ames: Iowa State College Film Production Unit, 1959.
Rudolf E. Schipull Papers.
David W. Staniforth Papers.
E. P. Sylwester Papers.

STATE HISTORICAL SOCIETY OF IOWA, IOWA CITY

Ray L. Gribben Manuscripts and Diary.
Flora and Ferd Jarrott Account Books.
Melvin Laughlin Papers.

STATE HISTORICAL SOCIETY OF IOWA, DES MOINES

Haight/Whittlesey Family Farm Records.
William Mason Family Farm Records.
Asa T. Meelchryst Papers.
Carl T. and Bertha Peterson Farm Records.

Periodicals

Agricultural Engineering
Ames Daily Tribune
Des Moines Register
Des Moines Tribune
The Farmer's Digest
Farm Implement News
Farm Industry News
Farm Journal
Farm Science Reporter
Fertilizer Review
Implement and Tractor
Iowa Agriculturist
Iowa Farm and Home Register
Iowa Farm Science
National Geographic
Successful Farming
Wallaces Farmer (Wallaces' Farmer and Iowa Homestead; Wallaces' Farmer)

Interviews

Anderson, R. D. Interview by author, 7 December 2001, Des Moines, Iowa. Tape recording.

Brubaker, Doyle. Interview by author, 7 September 2001, Newton, Iowa. Tape recording.

Hansen, Melvin (pseudonym). Interview by author, 9 June 2003, Madrid, Iowa. Tape recording.

Havran, Dennis. Interview by author, 11 January 2002, Des Moines, Iowa. Tape recording.

Johnson, J. Clifford. Interview by author, 14 December 2002, Council Bluffs, Iowa. Tape recording.

Johnson, Jim. Interview by author, 21 June 2004, Eldora, Iowa. Tape recording.

Meyer, Darlene and Elmer. Interview by author, 14 August 2001, Bridgewater, Iowa. Tape recording.

Neven, Gene. Interview by author, 12 October 2002, Marshalltown, Iowa. Tape recording.

Nymand, Robert. Interview by author, 30 October 2002, Brayton, Iowa. Tape recording.

Robinson, Keith. Interview by author, 1 October 1992, Atlantic, Iowa. Tape recording.

Sams, Clifford. Interview by author, 13 September 2002, Conrad, Iowa. Tape recording.

Smalley, Marsha and Ken. Interview by author, 21 June 2004, Iowa City, Iowa. Tape recording.

Westphalen, Johnnie. Interview by author, 2 October 1992, Atlantic, Iowa. Tape recording.

Government Documents and Publications

Anderson, M. A., L. E. Cairns, Earl O. Heady, and E. L. Baum. "An Appraisal of Factors Affecting the Acceptance and Use of Fertilizer in Iowa, 1953." *Special Report No. 16*. Agricultural Experiment Station, Iowa State College, Ames, June 1956.

Beal, George M., and Everett M. Rogers. "The Adoption of Two Farm Practices in a

Central Iowa Community." *Special Report No. 26.* Agricultural and Home Economics Experiment Station, Iowa State University, Ames, June 1960.

"Behavior Studies Related to Pesticides." *Special Report No. 49.* Agricultural and Home Economics Experiment Station, Cooperative Extension Service, Iowa State University, December 1966.

Bohlen, Joe M., George M. Beal, and Daryl Hobbs. "The Iowa Farmer and Farm Chemicals: Attitudes, Level of Knowledge and Patterns of Use." *Rural Sociology Report No. 3.* Department of Economics and Sociology, Iowa State University, Ames, November 1958.

Brodell, Albert P., and Martin R. Cooper. "The Costs and Ways of Making Hay." In *Grass: The Yearbook of Agriculture, 1948,* 173–77. Washington, DC: GPO, 1948.

Burnside, Orvin C., ed. *Biologic and Economic Assessment of Benefits from Use of Phenoxy Herbicides in the United States.* National Agricultural Pesticide Impact Assessment Program (NAPIAP), USDA, Special Report Number 1-PA-96. Washington, DC: GPO, November 1996.

———. "The History of 2,4-D and Its Impact on Development of the Discipline of Weed Science in the United States." In Burnside, *Phenoxy Herbicides,* 5–11.

Burroughs, Wise, C. C. Culbertson, and William Zmolek. "Questions and Answers about Adding Stilbestrol to Feeds for Growing and Fattening Beef Cattle." *Pamphlet 215.* Agricultural Extension Service, Iowa State College, Ames, November 1954.

Church, Lillian. "Partial History of the Development of Grain Harvesting Equipment." *Information Series No. 72.* Revised edition. Washington, DC: GPO, October 1947.

Cleaver, Thayer, and Robert G. Yeck. "Loose Housing for Dairy Cattle." In USDA, *Information Bulletin No. 98.* Washington, DC: GPO, 1953.

"Consumption of Commercial Fertilizers, Primary Plant Nutrients, and Micronutrients." In USDA Statistical Reporting Service, Crop Reporting Board, *Statistical Bulletin No. 472.* Washington, DC: GPO, 1971.

"Controlling the European Corn Borer." *Pamphlet 150.* Agricultural Extension Service, Iowa State College, Ames, December 1949.

"Cutting Costs in Today's Farming." *Pamphlet 222.* Agricultural Extension Service, Iowa State College, Ames, February 1956.

"Drying Ear Corn by Mechanical Ventilation." In USDA, *Miscellaneous Publication No. 919.* Washington, DC: GPO, 1963.

Eighth Biennial Report of Iowa Book of Agriculture, 1966–1967. Des Moines: State of Iowa, 1968.

Eleventh Biennial Report of the Iowa Department of Agriculture, 1971–1973. Des Moines: State of Iowa, 1973.

Environmental Protection Agency (EPA) Hazardous Materials Advisory Committee. *Herbicide Report: Chemistry and Analysis, Environmental Effects, Agricultural and Other Applied Uses.* Washington, DC: GPO, May 1974.

"The European Corn Borer and Its Control in the North Central States." *Pamphlet 176.* Cooperative Extension Service and Agricultural and Home Economics Experiment Station, Iowa State University, Ames. Revised October 1961, October 1968.

"Fertilizers and Lime in the United States: Resources, Production, Marketing, and Use." In USDA, *Miscellaneous Publication No. 586.* Washington, DC: GPO, May 1946.

"Fertilizers for Field Crops." *Pamphlet 112.* Agricultural Extension Service, Iowa State College, Ames. Revised January 1947.

"Fertilizer Use for Efficient Crop Production." *Pamphlet 227.* Cooperative Extension Service, Iowa State University, Ames. Revised March 1962.

Fifth Biennial Report of Iowa Book of Agriculture, 1960–1961. Des Moines: State of Iowa, 1962.

Fiftieth Annual Iowa Year Book of Agriculture. Des Moines: State of Iowa, 1949.

First Biennial Report of Iowa Book of Agriculture, 1952–1953. Des Moines, State of Iowa, 1954.

"Flies on Livestock." *Pamphlet 200.* Agricultural Extension Service, Iowa State College, Ames, May 1953.

Fourth Biennial Report of Iowa Book of Agriculture, 1958–1959. Des Moines: State of Iowa, 1960.

Grass: The Yearbook of Agriculture, 1948. Washington, DC: GPO, 1948.

Gray, R. B. "Equipment for Making Hay." In *Grass: The Yearbook of Agriculture, 1948,* 168–72. Washington, DC: GPO, 1948.

"Guide to Fertilizer Use." *Pamphlet 193.* Agricultural Extension Service, Iowa State College, Ames, March 1953.

"Guide to Higher Soybean Yields." *Pamphlet 202.* Agricultural Extension Service, Iowa State College, Ames, May 1953.

Gunderson, Harold. "Stop Corn Borer Damage!" *Pamphlet 134.* Agricultural Extension Service, Iowa State College, Ames, May 1948.

Gunderson, Harold, and J. H. Lilly. "Control Corn Rootworms." *Pamphlet 178.* Agricultural Extension Service, Iowa State College, Ames. Revised April 1953.

Hecht, Rueben W. "Labor Used for Livestock." In USDA Agricultural Research Service, *Statistical Bulletin No. 161.* Washington, DC: GPO, May 1955.

Hoiberg, Eric O., and Wallace Huffman. "Profile of Iowa Farms and Farm Families: 1976." *Bulletin P-141.* Agriculture and Home Economics Experiment Station and Cooperative Extension Service, Iowa State University, Ames, April 1978.

Holladay v. Chicago, Burlington & Quincy Railroad Company. 255 F. Supp. 879 (U.S. Dist. 1966).

"How Farm People Accept New Ideas." *Special Report No. 15.* Iowa State College, Ames, 1955.

Humphries, W. R., and R. B. Gray. "Partial History of Haying Equipment." In USDA, *Information Series No. 74.* Washington, DC: GPO. Revised October 1949.

Iowa State Department of Health. *Water Pollution Control Progress Report, 1968–1969.* Iowa Water Pollution Control Commission, Des Moines.

———. *Water Pollution Control Progress Report, July 1, 1969–June 30, 1970.* Iowa Water Pollution Control Commission, Des Moines.

———. *Water Pollution Control Progress Report, July 1, 1970–June 30, 1971.* Iowa Water Pollution Control Commission, Des Moines.

Johnson, Rebecca A., and Elizabeth V. Wattenberg. "Risk Assessment of Phenoxy Herbicides: An Overview of the Epidemiology and Toxicology Data." In Burnside, *Phenoxy Herbicides,* 16–40.

Kaldor, Donald R., and Earl O. Heady. "An Exploratory Study of Expectations, Uncertainty and Farm Plans in Southern Iowa Agriculture." *Research Bulletin 408.* Agricultural Experiment Station, Iowa State College, Ames, April 1954.

"Loose Housing for Dairy Cattle." In USDA, *Miscellaneous Publication No. 859.* Washington, DC: GPO, 1961.

"Making the Most from Fall-Applied Anhydrous Ammonia for Corn." *Pm-334.* Cooperative Extension Service, Iowa State University, Ames, August 1966.

Mitchell, John W. "Plant Growth Regulators." In *Science and Farming: The Yearbook of Agriculture, 1943–1947,* 256–66. Washington, DC: GPO, 1947.

Moe, Edward O., and Carl G. Taylor. "Culture of a Contemporary Rural Community, Irwin, Iowa." *Rural Life Studies* 5 (December 1942). Washington, DC: GPO.

Ninth Biennial Report of Iowa Book of Agriculture, 1968–1969. Des Moines: State of Iowa, 1970.

"Profitable Corn Production." *Pamphlet 409.* Cooperative Extension Service, Iowa State University, Ames, January 1968.

"A Quick Guide for Higher Soybean Yields." *Pamphlet 290.* Cooperative Extension Service, Iowa State University, Ames, May 1962.

Second Biennial Report of Iowa Year Book of Agriculture, 1954–1955. Des Moines: State of Iowa, 1956.

Seventh Biennial Report of Iowa Book of Agriculture, 1964–1965. Des Moines: State of Iowa, 1966.

"Storage of Small Grains and Shelled Corn on the Farm." In USDA, *Farmers' Bulletin No. 2009.* Washington, DC: GPO, 1949.

Sylwester, E. P., A. L. Bakke, and D. W. Staniforth. "Recommendations for Chemical Weed Control." *Pamphlet 140.* Agricultural Extension Service, Iowa State College, Ames, March 1949.

Tenth Biennial Report of Iowa Year Book of Agriculture, 1970–1971. Des Moines: State of Iowa, 1972.

Third Biennial Report of Iowa Year Book of Agriculture, 1956–1957. Des Moines: State of Iowa, 1958.

U.S. Census of Agriculture, 1940. Vol. 1, part 2. Washington, DC: GPO, 1942.

U.S. Census of Agriculture, 1950. Vol. 1, part 2. Washington, DC: GPO, 1952.

U.S. Census of Population: 1960. Vol. 1, *Characteristics of the Population, part 17: Iowa.* Washington, DC: GPO, 1963.

U.S. Department of Agriculture (USDA). *Agricultural Statistics, 1950.* Washington, DC: GPO, 1950.

"Weed Control in Corn." *Pamphlet 269.* Iowa State University Extension, Ames, March 1964.

"Weed Control in Soybeans." *Pamphlet 270.* Iowa State University Extension, Ames, April 1969.

"We Must Have More Soybeans and Flax!" *Pamphlet 26.* Agricultural Extension Service, Iowa State College, Ames, March 1942.

Zmolek, William, and Wise Burroughs. "62 Questions about Stilbestrol Answered." *Bulletin P-133.* Agricultural and Home Economics Experiment Station, Cooperative Extension Service, Iowa State University, Ames, July 1963.

Trade Literature and Promotional Films

Into Tomorrow. 8mm. Produced by the Calvin Company for Massey-Harris Company, directed by Reese Wade, 1946.

"John Deere No. 12-A Combine Paves the Way to Bigger Profits." John Deere Sales Brochure, A-534-50-8, 1950.

"KX Research at Work for You." KX-57-4. Northrup, King & Company, 1957.

The Sheppards Take a Vacation. 8mm. Produced by Ray-Bell Films for John Deere Company, circa 1940.

Secondary Sources

Aldrich, Samuel R., and Earl R. Leng. *Modern Corn Production.* Cincinnati, OH: Farm Quarterly, 1965.

Bauerkamper, Arnd. "The Industrialization of Agriculture and Its Consequences for the Natural Environment: An Inter-German Comparative Perspective." *Historical Social Research* 29.3 (2004): 124–49.

Bear, Firman E. *Soils and Fertilizers.* 4th ed. New York: John Wiley and Sons, 1951.

Beeman, Randal S., and James A. Pritchard. *A Green and Permanent Land: Ecology and Agriculture in the Twentieth Century.* Lawrence: University Press of Kansas, 2001.

Bogue, Allan G. "Changes in Mechanical and Plant Technology: The Corn Belt, 1910–1940." *Journal of Economic History* 43 (March 1983): 1–25.

———. *From Prairie to Corn Belt: Farming on the Illinois and Iowa Prairies in the Nineteenth Century*. Chicago: University of Chicago Press, 1963.

Bohlen, Joe M. "Adoption and Diffusion of Ideas in Agriculture." In *Our Changing Rural Society,* ed. James H. Copp, 265–87. Ames: Iowa State University Press, 1964.

Brandsberg, George. *The Two Sides in the NFO's Battle.* Ames: Iowa State University Press, 1964.

Bray, James O., and Patricia Watkins. "Technical Change in Corn Production in the United States, 1870–1960." *Journal of Farm Economics* 46 (November 1964): 751–65.

Bremer, Richard G. *Agricultural Change in an Urban Age: The Loup Country of Nebraska, 1910–1970.* Lincoln: University of Nebraska Press, 1976.

Browning, George M. "Agricultural Pollution—Sources and Control." In *Water Pollution Control and Abatement,* ed. Ted L. Willrich and N. William Hines, 150–66. Ames: Iowa State University Press, 1967.

Buhs, Joshua Blu. *The Fire Ant Wars: Nature, Science, and Public Policy in Twentieth Century America.* Chicago: University of Chicago Press, 2004.

Cochrane, Willard W. *The Development of American Agriculture: A Historical Analysis.* Minneapolis: University of Minnesota Press, 1979.

Cochrane, Willard W., and Mary E. Ryan. *American Farm Policy, 1948–1973.* Minneapolis: University of Minnesota Press, 1981.

Colbert, Thomas Burnell. "Iowa Farmers and Mechanical Corn Pickers, 1900–1952." *Agricultural History* 74 (Spring 2000): 530–44.

Daniel, Pete. *Breaking the Land: The Transformation of Cotton, Tobacco, and Rice Cultures since 1880.* Urbana: University of Illinois Press, 1985.

———. *Lost Revolutions: The South in the 1950s.* Chapel Hill: University of North Carolina Press, 2000.

Fiege, Mark. *Irrigated Eden: The Making of an Agricultural Landscape in the American West.* Seattle: University of Washington Press, 1999.

Fink, Deborah, *Open Country: Rural Women, Tradition, and Change.* Albany: State University of New York Press, 1986.

Finke, Jeffrey. "Nitrogen Fertilizer: Price Levels and Sales in Illinois, 1945–1971." *Illinois Agricultural Economics* 13 (January 1973): 34–40.

Finlay, Mark R. "Hogs, Antibiotics, and the Industrial Environments of Postwar Agriculture." In *Industrializing Organisms: Introducing Evolutionary History,* ed. Susan R. Schrepfer and Philip Scranton, 237–60. New York and London: Routledge, 2004.

Fite, Gilbert C. *American Farmers: The New Minority.* Bloomington: Indiana University Press, 1981.

———. "The Transformation of South Dakota Agriculture: The Effects of Mechanization, 1939–1964." *South Dakota History* 19 (Fall 1989): 278–305.

Fitzgerald, Deborah. "Beyond Tractors: The History of Technology in American Agriculture." *Technology and Culture* 32 (January 1991): 114–26.

———. *Every Farm a Factory: The Industrial Ideal in American Agriculture.* New Haven: Yale University Press, 2003.

Gardner, Bruce L. *American Agriculture in the Twentieth Century: How It Flourished and What It Cost.* Cambridge: Harvard University Press, 2002.

Giglio, James N. "New Frontier Agricultural Policy: The Commodity Side, 1961–1963." *Agricultural History* 61 (Summer 1987): 53–70.

Grant, H. Roger, and Edward Purcell, eds. *Years of Struggle: The Farm Diary of Elmer G. Powers.* Ames: Iowa State University Press, 1976.

Hallberg, Milton C. *Economic Trends in U.S. Agriculture and Food Systems since World War II.* Ames: Iowa State University Press, 2001.

Hamilton, Carl. *In No Time At All.* Ames: Iowa State University Press, 1974.

Hansen, John Mark. *Gaining Access: Congress and the Farm Lobby, 1919–1981*. Chicago: University of Chicago Press, 1991.

Hart, John Fraser. "Part-Ownership and Farm Enlargement in the Midwest." In *Annals of the Association of American Geographers* 81 (March 1991): 66–79.

Hays, Samuel P. *Beauty, Health, and Permanence: Environmental Politics in the United States, 1955–1985*. Cambridge: Cambridge University Press, 1989.

Haystead, Ladd, and Gilbert C. Fite. *The Agricultural Regions of the United States*. Norman: University of Oklahoma Press, 1955.

Hilliard, Jules B. "The Revolution in American Agriculture." *National Geographic* (February 1970): 147–85.

Hilliard, Sam Bowers. "The Dynamics of Power: Recent Trends in Mechanization on the American Farm." *Technology and Culture* 13 (January 1972): 1–24.

Hines, N. William. *Nor Any Drop to Drink: Public Regulation of Water Quality*. Iowa City: Agricultural Law Center, College of Law, University of Iowa, September 1967.

Holm, Lennis J. "From Family Farm to Agribusiness." Honors thesis, University of Iowa, 1967.

Hudson, John. *Making the Corn Belt: A Geographical History of Middle-Western Agriculture*. Bloomington and Indianapolis: Indiana University Press, 1994.

Hurt, R. Douglas. *Agricultural Technology in the Twentieth Century*. Manhattan, KS: Sunflower University Press, 1991.

———. *American Farm Tools from Hand Power to Steam Power*. Manhattan, KS: Sunflower University Press, 1986.

———. "Ohio Agriculture since World War II." *Ohio History* 97 (Winter–Spring 1988): 50–71.

———. *Problems of Plenty: The American Farmer in the Twentieth Century*. Chicago: Ivan R. Dee, 2002.

Jellison, Katherine. *Entitled to Power: Farm Women and Technology, 1913–1963*. Chapel Hill: University of North Carolina Press, 1993.

Kline, Gerald L. "Harvesting Hay with the Automatic Field Baler." Master's thesis, Iowa State College, 1946.

Kline, Ronald. *Consumers in the Country: Technology and Social Change in Rural America*. Baltimore: Johns Hopkins University Press, 2000.

Marcus, Alan I. *Cancer from Beef: DES, Federal Food Regulation, and Consumer Confidence*. Baltimore: Johns Hopkins University Press, 1994.

Marcus, Alan I, and Howard P. Segal. *Technology in America: A Brief History*. San Diego: Harcourt, Brace, Jovanovich, 1989.

Matusow, Allen J. *Farm Policies and Politics in the Truman Years*. Cambridge: Harvard University Press, 1967.

Mayda, Chris. "Pig Pens, Hog Houses, and Manure Pits: A Century of Change in Hog Production." *Material Culture* 36 (Spring 2004): 18–42.

McMath, Robert. "Where's the 'Culture' in Agricultural Technology?" In *Outstanding in His Field: Perspectives on American Agriculture in Honor of Wayne D. Rasmussen*, ed. Frederick V. Carstensen, Morton Rothstein, and Joseph A. Swanson, 123–37. Ames: Iowa State University Press, 1993.

Midwest Farm Handbook. Fifth edition. Ames: Iowa State University Press, 1960.

Morris, Robert L., and Lauren G. Johnson. "Pollution Problems in Iowa." In *Water Resources of Iowa*, ed. Paul J. Horick, 89–109. Iowa City, University Printing Service, 1970.

Muhm, Don. *The NFO: A Farm Belt Rebel: History of the National Farmers Organization*. Red Wing, MN: Lone Oak Press, 2000.

National Agricultural Statistics Service, USDA. www.nass.usda.gov?ia/historic/crn1866.txt. Accessed 17 May 2005.

Nelson, Philip J. "To Hold the Land: Soil Erosion, Agricultural Scientists, and the Development of Conservation Tillage Techniques." *Agricultural History* 71 (Winter 1997): 71–90.

Neth, Mary. *Preserving the Family Farm: Women, Community, and the Foundations of Agribusiness in the Midwest, 1900–1940.* Baltimore: Johns Hopkins University Press, 1995.

Oudshoorn, Nelly, and Trevor Pitch. *How Users Matter: The Construction of Users and Technologies.* Cambridge: MIT Press, 2003.

Page, William Colgan. *Leaner Pork for a Healthier America: Looking Back on the Northeast Iowa Swine Testing Station.* Des Moines: Iowa Department of Transportation in cooperation with the Federal Highway Administration and the State Historical Society of Iowa, 2000.

Peoples, Kenneth L., David Freshwater, Gregory D. Hanson, Paul T. Prentice, and Eric P. Thor. *Anatomy of an American Agricultural Credit Crisis: Farm Debt in the 1980s.* Lanham, MD: Rowman and Littlefield, 1992.

Petersen, Gale. "The Discovery and Development of 2,4-D." *Agricultural History* 41 (July 1967): 243–53.

Quick, Graeme, and Wesley Buchele. *The Grain Harvesters.* St. Joseph, MI: American Society of Agricultural Engineers, 1978.

Rasmussen, Nicolas. "Plant Hormones in War and Peace: Science, Industry, and Government in the Development of Herbicides in 1940s America." *Isis* 92 (June 2001): 291–316.

Rasmussen, Wayne D. "The Impact of Technological Change on American Agriculture, 1862–1962." *Journal of Economic History* 4 (December 1962): 578–91.

———. *Taking the University to the People: Seventy-Five Years of Cooperative Extension.* Ames: Iowa State University Press, 1989.

Rikoon, J. Sanford. *Threshing in the Midwest: A Study of Traditional Culture and Technological Change, 1820–1940.* Bloomington and Indianapolis: Indiana University Press, 1988.

Riney-Kehrberg, Pamela. *Rooted in Dust: Surviving Drought and Depression in Southwestern Kansas.* Lawrence: University Press of Kansas, 1994.

Robbins, Paul R. "Labor Situations Facing the Producer." In *The Pork Industry: Problems and Progress,* ed. David G. Topel, 211–19. Ames: Iowa State University Press, 1968.

Ross, Earle D. *Iowa Agriculture.* Iowa City: State Historical Society of Iowa Press, 1951.

Rothman, Hal. *Saving the Planet: The American Response to the Environment in the Twentieth Century.* Chicago: Ivan R. Dee, 2000.

Russell, Edmund. *War and Nature: Fighting Humans and Insects from World War I to Silent Spring.* Chapel Hill: University of North Carolina Press, 2001.

Sawyer, Richard C. "Monopolizing the Insect Trade: Biological Control in the USDA, 1888–1951." In *The United States Department of Agriculture in Historical Perspective,* ed. Alan I Marcus and Richard Lowitt, 271–85. Washington, DC: Agricultural History Society, 1991.

Schlebecker, John T. *A History of American Dairying.* Chicago: Rand McNally, 1967.

———. *Whereby We Thrive: A History of American Farming, 1607–1972.* Ames: Iowa State University Press, 1975.

Schmitt, Richard George, Jr. "Economic Analysis of Haying Methods in Eastern Iowa." Master's thesis, Iowa State College, 1947.

Schnapsmeier, Edward L., and Frederick H. Schnapsmeier. "Eisenhower and Ezra Taft Benson: Farm Policy in the 1950s." *Agricultural History* 44 (October 1970): 369–78.

Schwieder, Dorothy. *75 Years of Service: Cooperative Extension in Iowa.* Ames: Iowa State University Press, 1993.

Shover, John. *First Majority, Last Minority: The Transformation of Rural Life in America.* DeKalb: Northern Illinois University Press, 1977.

Soike, Lowell, "Viewing Iowa Farmsteads." In *Take This Exit: Rediscovering the Iowa Landscape,* ed. Robert F. Sayre, 153–72. Ames: Iowa State University Press, 1989.

Sprague, G. F., and J. C. Cunningham. "Growing the Bumper Corn Crop." In *A Century of Farming in Iowa, 1846–1946,* 32–44. Ames: Iowa State College Press, 1946.

Summons, Terry G. "Animal Feed Additives, 1940–1966." *Agricultural History* 42 (October 1968): 305–13.

Taylor, C. Robert. "An Analysis of Nitrate Concentrations in Illinois Streams." *Illinois Agricultural Economics* 13 (January 1973): 12–19.

Taylor, Fred W. "North Dakota Agriculture since World War II." *North Dakota History* (April 1967): 47–61.

Wendel, C. H. *The Allis-Chalmers Story.* Osceola, WI: Crestline Publishing, 1993.

"We've Come a Long Way." 2, 4-D Skit, www.weeds.iastate.edu/weednews/24dplay.htm. Accessed 12 November 2007.

Wilcox, Walter W. *The Farmer in the Second World War.* Ames: Iowa State College Press, 1947.

Young, James A. "The Public Response to the Catastrophic Spread of Russian Thistle (1880) and Halogeton (1945)." In *Publicly Sponsored Agricultural Research in the United States: Past, Present, and Future,* ed. David B. Danbom, 122–30. Washington, DC: Agricultural History Society, 1988.

Zimmerman, O. T., and Irvin Lavine. *DDT: Killer of Killers.* Dover, NH: Industrial Research Service, 1946.

Index